DON'T BUILD, REBUILD

ALSO BY AARON BETSKY

The Monster Leviathan: Anarchitecture

50 Lessons to Learn from Frank Lloyd Wright
(with Gideon Fink Shapiro)

Renny Ramakers: Rethinking Design

Architecture Matters

Making It Modern: The History of Modernism in Architecture of Design

At Home in Sprawl

The U.N. Building

False Flat: Why Dutch Design Is So Good
(with Adam Eeuwens)

Landscrapers: Building with the Land

Three California Houses: The Homes of Max Palevsky

Architecture Must Burn (with Erik Adigard)

Queer Space: Architecture and Same-Sex Desire

Building Sex: Men, Women, Architecture, and the Construction of Sexuality

James Gamble Rogers and the Architecture of Pragmatism

Experimental Architecture in Los Angeles
(with John Chase and Leon Whiteson)

Violated Perfection

DON'T BUILD, REBUILD

THE CASE FOR IMAGINATIVE REUSE IN ARCHITECTURE

AARON BETSKY

BEACON PRESS, BOSTON

BEACON PRESS
Boston, Massachusetts
www.beacon.org

Beacon Press books
are published under the auspices of
the Unitarian Universalist Association of Congregations.

Printed in the United States of America

27 26 25 24 8 7 6 5 4 3 2 1

This book is printed on acid-free paper that meets the uncoated paper
ANSI/NISO specifications for permanence as revised in 1992.

Text design and composition by Kim Arney

Library of Congress Cataloguing-in-Publication
Data is available for this title.
Hardcover ISBN: 978-0-8070-1486-8
E-book ISBN: 978-0-8070-1487-5
Audiobook: 978-0-8070-1720-3

To those who reuse

CONTENTS

DON'T BUILD, REBUILD

INTRODUCTION

THE PRINCIPLES OF IMAGINATIVE REUSE

We should not construct new buildings from scratch. That is the simple message of this book. I do not make this statement lightly. I have spent my life looking at, evaluating, loving, and even trying to make my own architecture. And having grown up in the Netherlands, where my family moved when I was a young child, I appreciate the ways in which architecture can build a new and better world. The Dutch have been doing that for centuries, creating a fully formed country out of the sea and swamps.

Throughout my career as a designer and critic, I have traveled around the globe and seen many fantastic buildings, new and old. In recent years, however, I found myself attracted to a new type of architecture: one that does not impose itself on the land as new but that reveals, opens, and reuses the world we have inherited in an imaginative manner. I believe there are reasons for that beauty, and I have spent the last few years seeking out and analyzing such structures.

I have concluded that there are two reasons to commit to an architecture of reuse. The first is that every act of new construction—no matter how carefully thought-out and how overlaid with solar panels, double or even triple-paned windows, rainwater collectors, geothermal wells, ceiling fans, or other gadgets—will only contribute to climate change. While existing buildings are carbon sinks, new buildings use up resources we cannot replenish. To build with new steel, you must mine ore, smelt it, and transport the results. Concrete production is even worse as an extractive, heat-producing, and chemical-using industry. Even wood either uses forests that should be sucking up carbon (and just being) or facilitates the unhealthy monoculture of fast-growing tree plantations. The chemicals we use to create coverings for walls, ceilings, and fire protections are almost all petroleum based. These concerns do not even include all the other materials we use to make buildings and furnish them, let alone the transportation of all of this to the site.

Architects can claim that their buildings are "net zero"—that is, produce no net pollution because they use alternative energy sources such as solar panels or geothermal heating and cooling—or even "net positive," but in fact the energy debt they build up in their construction is just about impossible to repay. The reuse of existing structures is the only way to begin to approach the need to preserve our wood, sand, silicon, and iron, let alone stop belching carbon monoxide and other pollutants in the air.

The second reason to embrace reuse is that any new building will by its very nature reflect the values, beliefs, and mandates of those who commission it, which is to say people with means. We need to stop ruining our planet, and we also need to open the built spaces of our society to be more egalitarian and just.

Back in 2008, the European Union proposed that 70 percent of all buildings be constructed out of reused or recycled material by 2020.[1] They also looked to cut by half the amount of construction waste that would wind up in landfills and suggested other ways in which the whole industry should be more sustainable in its practices. While the US made no such proposal, American architecture has also been moving toward reuse: in 2022, more than half of firms' billings were for renovation work of various sorts. That milestone was reached long before in Europe—where renovation and restoration have been an overwhelming challenge since the devastation of the Second World War—and many countries in Asia may soon hit this mark as well.

The economic factors driving us toward reuse are availability and cost. Because of the dearth of both land and raw materials—as well as the regulations we put on their use—the costs of new construction have risen to such a degree that they often outweigh the added complexity and expense that go into renovating existing buildings. This is by no means always the case, however, and most developers—and architects—will continue to make new buildings. That is not only because it is cheaper in many cases but also because it is what they know how to do, and they believe it to be the highest and best way to apply their skills and knowledge. An essential part of the process of revaluing existing buildings will indeed have to be changes to the culture of architecture—and the culture of the arts in general—that favor the creative reuse of materials, forms, and buildings over the production of brand-new buildings and other cultural objects.

Since the onset of the COVID pandemic and the shift to remote work, we have seen a wave of office building conversions, but in many cases there is a great deal of resistance to these conversions because of the manner in which building codes are written. Most cities restrict the uses to which buildings can be put in certain areas ("zoning") and have very specific regulations about matters such as how many exit stairs and bathrooms, and even parking spaces, you need for each building type. When you want to change the use of a building from workplace to residential, you not only have to often appeal for a variance for the use designation but also add stairs, individual bathrooms (while removing the communal ones of offices), and sometimes even change materials.

Right now, most companies and services that offer redesign services and goods, whether in design or by providing upcycled materials, are small and often experimental. As we get better at reusing materials and adapting buildings, and as we rewrite our laws to accommodate such changes, this form of reuse will also become more economical. A rise in demand for redesigned goods and services should create more knowledge and increase the scale of these operations.

The stick of standards that governments are setting for circular economies of reuse in Europe and—to a lesser degree, in the United States and elsewhere—will boost the carrot of making reuse cheaper. The idea of a circular economy, which started in the 1970s with regulations on the recycling of paper and plastic, is now reaching into the building industry as well.

THE SOCIAL ROLE OF ARCHITECTURE

If we were to merely stop building altogether, we would make great progress toward a sustainable and equitable world. Yet we would also lose the contributions that architecture makes to our culture and society. Architecture can be a way of setting a scene in which we can play the role we believe is right for us, in relationship to both the human-made and the natural worlds. It can open our eyes to new realities and let us rediscover where we have come from. Our houses are domestic scenes, and our offices are stages where we perform the roles of workers, but how they do that is predetermined and imposed on us, even if we arrange our furniture or our desk to make it our own. An architecture of reuse needs to make these places better. Architects such as Frank Gehry have long been experimenting with bare-bones structures (in his work of the 1970s and 1980s, Gehry often left the wood studs of buildings exposed), so that they are more open for use and interpretation. Designers such as Rem Koolhaas and the Dutch firm MVRDV have designed buildings in which work spaces double as stages—stepped seating and corridors spiral into continual paths of discovery while widening at times to encourage flexible uses, including social and commercial, for example.

We need architecture to make our world better, but, as it turns out, architecture does not need to make new buildings to do so. It can reuse, re-inhabit, and reimagine the structures we already have. It can preserve and open up at the same time. It can find beauty where we thought there was only the ugly and humdrum, and it can surround us with what it has discovered. This is the kind of reuse—what I am calling "imaginative reuse"—in which we must engage. This new way of reuse has arisen quite recently, in just the last few years—although its roots, as this book will try to show, reach back centuries and even millennia. Rather than accepting or trying to rebuild history, imaginative reuse treats the past as a continuous process that continues to this day and will keep evolving in the future. Buildings keep changing as they exist, and so do the lives within them. Whatever we do in or with those buildings, we should understand our actions as part of this process. Imaginative reuse should reveal the past, much as restorers have often tried to do, but it should also develop a critical attitude toward that past and elaborate its possibilities. It should accept wear and tear, fragments, and even ruins, and both re-inhabit and reimagine them. What we need, in other words, is to move beyond the five traditional categories that have up until now defined the reuse of buildings.

TRADITIONAL CATEGORIES OF REUSE

The first category, *simple reuse*, has been practiced almost since the first buildings were erected: this is the act of occupying spaces with whatever function was needed at the time, with little regard for intentions of the original building. The second category, *reuse of materials*, involves taking buildings apart down to their base components and making these *spolia* (the Latin word for the reuse of piece of old buildings in new ones) part of our new structures—again often with little regard for how these stones, bricks, pieces of wood, and columns had functioned originally.

Starting at the end of eighteenth century—though of course you can find much earlier examples throughout history—people became interested in a third category of reuse: *preserving artifacts* we felt were important, whether because of their association with important events and people or because of their intrinsic beauty. The French Revolution swept away the old order and with it the need for palaces and churches, but the revolution also encouraged their reuse. Thus the Louvre, once the French king's home, became a public museum where the royal collections are now visible to everybody. The same events prompted both scholars and politicians, often for different reasons, to preserve and even resurrect fragments of the old order. In England, poets bemoaned the loss of countryside and innocence by trekking to ruins such as Tintern Abbey (as William Wordsworth did), while in Germany painters such as David Caspar Friedrich used similar crumbling remains of old churches to reflect a romantic desire to muse on time and place. At the same time, the rising cult of the genius led to a desire to preserve places where great men (almost never women) had lived or acted.

This way of saving the past allows it to be present in our current lives. The preservation movement started with such monuments, but since then has spread in many directions. Whole districts—from Manhattan's bustling Upper East Side to Cincinnati's Over-the-Rhine, which contains the largest collection of Victorian-era buildings in America—are now considered historic monuments that must be preserved as if in amber. This attitude, however, has little regard for how we live, work, and play today or how we will tomorrow. It seeks to make the past—or at least a version of it embodied in a building—present.

As the preservation movement spread beyond monuments, it intersected with the everyday practice of reusing old buildings, and a fourth category of reuse developed: *adaptive reuse*. This is currently our most common approach to the structures that we inherit and do not tear down: We make them work for us, with various degrees of changes, additions, and subtractions. We turn Washington, DC's grand former post office into a hotel, cramming bedrooms into what had been suites of offices and laying out seas of dining tables and bar stools in what had been the sky-lit sorting room. Or, in more humble examples, we turn a smaller post office in New York City's Soho district into a sleek new Apple store.

Finally, we should include in the panoply of different forms of reuse a fifth category, which encapsulates the way humans made ad hoc shelters before we became interested in building for the ages: the continual reassembly and use of materials that were flexible and easily deployable. We have done this since the very dawn of architecture, when we wove together tents out of wool or twigs, sometimes carrying them with us. Armies brought their tents with them. Seasonal fairs were made up of lean-tos and temporary pavilions. Today, such ephemeral architecture can be seen in the city built in a month for the Kumbh Mela festival in India, and in the instant cities erected for music festivals every summer.

All of these forms of reuse are now part of how we build and rebuild today to a greater or lesser degree, and they all are important in our quest to fully utilize what we already have. What imaginative reuse adds to these categories, I believe, is the framework of pursuing these forms of reuse as architecture. We need to continue to perfect the technical issues that allow us to improve our ability to reuse, while also changing our laws and regulations to make it easier to do so. We are getting better and better at standardizing the ways we sort and make available salvaged building components, using algorithms and web advertising to do so. In 2023 the pressure to ease the conversion of buildings from one use to another led New York City to change some of its regulations in this area.

COMBINING OLD AND NEW

But there is a deeper set of issues that architecture needs to pursue so that we may reuse our buildings in a better manner. Most of those conundrums form a simple contradiction: we want new things and an always newer world, but we want to be comfortable in that novelty. We want progress, but we also want to be at home in a world that is continually changing. It is a contradiction inherent to our capitalist economic system and the culture based on it. In our current logic of production and consumption, we depend on making and using ever more in ever more efficient ways to keep both the economy and profit margins growing (and sometimes costs going down). We are therefore conditioned by our schooling and advertising to always want newer and better products, even as we try to hold onto the memories, objects, and places that define us. Architecture therefore must be able to reuse buildings in a way that makes them feel new, but at the same time lets us feel as if we are sheltered by them, rooted in their foundations, comfortable in their materials.

Moreover, architecture has to confront the reality that its task is to open and preserve at the same time. Imaginative reuse must challenge the power relations installed in buildings by the wealthy people who commissioned them, yet also preserve the buildings materially so that they can be open to more people. The act of reuse should break through the hierarchies of class, sex, and race.

What we are looking for is perhaps a bit of a Frankenstein's monster, as the historian Liliane Wong has pointed out. We want the reused building to reflect us, and at the same time we realize a building is a modern machine. The eeriness of the combination is frightening. Wong sees the monster as analogous to a building that has been renovated, but in a way that is not neutral: "The substitution of different body parts within this structure, however, is a deviation, a subversion of the structure."[2]

The sense that a building is recognizable but alien to us can be resolved, Wong claims, if we realize that the act of reuse is also akin to an even older myth: that of resurrection. "In this respect, and in an interesting parallel to restoration, the aim of preservation until the early 20th century remained somewhat aligned with Christian resurrection, akin to Jesus' resurrection from death in perpetuity at age 33," she writes.[3] Perhaps the Christian reference is no longer directly present in the motivation for reuse, but what does or should remain is the sense that we are looking for an animating presence that resembles us, is also beyond us, and gives meaning to what we are doing.

A less laden way to say this is that while reuse always has a nostalgic side, we need to find ways not to wallow in those memories. Instead, we must activate them and make them part of how we move forward. The artist Svetlana Boym calls this "reflective nostalgia," which, she says, has "a utopian dimension that consists in the exploration of other potentialities and unfulfilled promises of modern happiness." That gives the ruin a redeeming quality: "It resists both the total reconstruction of the local culture and the triumphant indifference of technocratic globalism . . . the reflective nostalgics [sic] can create a global diasporic solidarity based on the experience of immigration and internal multiculturalism."[4]

The architects David Fannon, Michelle Laboy, and Peter Wiederspahn, in their 2021 *The Architecture of Persistence: Designing for Future Use*, relate this idea of reflective nostalgia (though they do not use that term) directly to the act of adaptive reuse. They point out that "material decay fuels the work of artists and architects because it imbues buildings with the passage of time." In response to that, they argue, "a pragmatic approach to an architecture of persistence recognizes the emotional and creative power of allowing traces of time to be registered in buildings while negotiating the inherent risks, technical dimensions, and performance criteria for contemporary material culture." For the authors, "This is an argument for embracing the role of materiality in the cultural persistence of buildings, to guard against unnecessarily accelerating the path to decay or demolition long before a building's material life span."[5]

"Keeping a building," they go on to say, "is an act of optimism. The premise is that the building is valuable enough to make it worth the effort to leave a new mark, to become part of the continuum of history while pushing a new generation forward."[6]

ARCHITECTURE AS ARCHAEOLOGY

How, then, do we pursue such a path? Ironically, it starts with doing nothing, or at least as little as we can. The Dutch architect Rem Koolhaas, writing about historic preservation, has said: "Perhaps in architecture, a profession that fundamentally is supposed to change things it encounters (usually before reflection), there ought to be an equally important arm of it that is concerned with not doing anything."[7]

We may not even have to choose to do nothing, if for no other reason that we can't do anything, he goes on to write: "We are living in an incredibly exciting and slightly absurd moment, namely that preservation is overtaking us. Maybe we can be the first to experience the moment that preservation is no longer a retroactive activity but becomes a prospective activity."[8]

That Zen-like act of doing nothing leads to particular strategies. The first of these is to see imaginative reuse as, above all else, a form of revelation—a kind of archaeology. If we treat buildings as excavation sites, our work consists primarily of revealing what is already there. Like archaeologists, we must be careful in that effort, but we also must accept what we find in all its variety, realizing that much of it will be remnants rather than complete structures. Moreover, an archaeological dig usually reveals many layers of construction from different times. Figuring out, first, how to keep what we find intact and, second, what to reveal becomes just as important as the act of opening up and removing the dirt between the ruins. When we preserve and renovate a historic house, do we take it back to the day the original resident died, or do we allow all the lives that existed there to also be present in their marks? At Taliesin West—the home, office, and school that Frank Lloyd Wright built for himself in 1937 in the Arizona desert—the foundation that purports to carry on his name is restoring the building to the way it supposedly looked the day he died in 1959, ripping our decades of additions Wright's disciples subsequently built. The artist David Ireland, on the other hand, as we shall see in chapter 4, took great pains to preserve as many marks as possible of the different lives that had taken place in the house he bought in San Francisco in 1974, weaving his own insertions into the existing structure. On a larger scale, the architect David Chipperfield left in place the bullet holes and craters left over from the Second World War that marred Berlin's Neues Museum when he renovated that structure in 2009.

Before we consider how to reuse, then, we must first ask ourselves: how do we activate these ruins to their fullest potential? Our task is to enable others to perceive the full reality of what we have found in all its many layers and potential imagery.

ARCHITECTURE AS ART

This is where architecture can learn a great deal from art practices that make it their business to evoke other worlds and times in ways that are convincing and, even more than that, seductive and eerie. Artists such as Mike Nelson create installations that seem to be office environments or even whole apartment buildings from an older New York or faraway Istanbul. Similarly, the Berlin-based artist duo Elmgreen & Dragset create stage sets in which you can imagine yourself wandering through the apartment of a recently deceased architect, calling into question the reality of what you are experiencing. Architecture has the advantage of making present the past in a manner that confronts you directly but that can also make it difficult to see what is within its surfaces and structures. The tactics of framing, evoking, collaging, and assembling images—sometimes of a three-dimensional nature, sometimes unreal or projected—show us how we can perform a more revelatory form of archaeology. As we shall see later in this book, these tactics were pioneered by artists such as Kurt Schwitters at the beginning of the twentieth century and can be fruitfully adapted by architects.

AN ARCHITECTURE OF DOING SOMETHING

Even as we excavate what is there, we usually still have to do something: we have to subtract from, add to, or multiply the existing structure (that is, weave the new through the old) in order to bring it into a state that allows present and future uses. Of these, the subtracting is easiest. Cutting open buildings in such a manner that we can utilize them is a necessity. The work of the architect and artist Gordon Matta-Clark, who sawed open apartment buildings, houses, and warehouses as art projects during the 1960s, showed the way for this technique. In Madrid, the architecture firm Herzog & de Meuron jacked up an existing warehouse on top of a new, two-level foundation so that you can pass into the open space of a new museum they created in the structure underneath the imposing walls. In Charleroi, Belgium, De Vylder Vinck Architects removed most of the walls of the central pavilion of a former exposition center to make it a public space. To create the Ponce City Market in Atlanta in 2018, S9 Architecture created large holes in what had been warehouse floors to create airy double-height public spaces on the ground level. Two years before, Flores & Prats made more delicate incisions in a two-story nineteenth-century social hall to make its center a gathering spot from which visitors can see different activities on various floors. Renovating buildings so that they are more open, both visually and socially, can thus turn that need for functional space into architecture.

Adding is more difficult. The standard in historic preservation is that what is added should be easily identifiable as such. If it is new, it should look new. It should also be easy to remove without damaging the historic structure. Even though there are no such rules for reusing non-protected buildings, it is a good strategy that is used by many, if not most, architects working on historic structures.

There is inevitably a style to this kind of addition, made out of both necessity and choice. These interventions should be minimal and minimalistic. This is because technology has become much more compressed and ergonomically housed—with formerly room-sized computers turning into handheld devices, and chairs streamlined to fit our bodies better—and also because such minimal forms will allow the past to remain while carving out a place for the present. Moreover, we should aspire to make these additions in a way that uses as few natural resources as possible. When possible, we should make use of recycled or upcycled materials. In addition, we should consider minimal climate conditioning in the manner of the LocHal, a former tram repair facility in Tilburg, Netherlands, that is now a library and community center.

Multiplication, or the weaving of the new through the old, is an even more difficult tactic. The LocHal does a good job of that, but you can also find it in bookstores that are inserted into churches (Maastricht, Netherlands) or in old theaters (Buenos Aires, Argentina). In contrast to the method of addition I discussed above, multiplication can mean deliberately blurring the boundaries between new and old. Though that would seem to be a recipe for confusion, the raising of doubts about what or from when a piece of construction originates can be a very productive tool for triggering our wonderment and thus awareness. In such a case, it is exactly the continuity of time and place that is at the heart of architecture that thus comes to light. Unlike other forms of art, buildings surround us with the past, even embodying it and making us aware that we exist in a social continuum. That can make buildings deadening and oppressive—as when our apartments dictate which space is a bedroom and which a kitchen—or, on a larger scale, can make then intimidating, as with imposing museum buildings like the Metropolitan in New York. This is why blurring the distinction between new and old makes it possible for us to see the multiple histories and their marks all around us. Multiplication is a way of opening up buildings without destroying them. It is also a kind of questioning of the closed nature of styles, building practices, and all the other parts of the architecture canon by showing how changeable they are.

Blurring is central to many of these approaches. Paradoxically, doing almost nothing, but fudging the hard lines between past, present, and possible future along the way—like Frank Gehry's transformation of another tram repair shop into the Temporary Contemporary in Los Angeles in 1982—releases them from their monumental status. It also lets us see what we are doing now in terms of continuities rather than breaks. When something is blurred, it has an evocative quality. This is an art of intimation, rather than imitation. It operates at best out of the corner of our eye, when we realize that what seems like simple structures are actually not so simple on closer look. Architecture that is fake but real, that is "both/and and neither/nor" (to use the postmodern architect Robert Venturi's phrase)[9] is precisely what we need in this uncertain age. It grounds us without burying us and invites us to interpret, speculate, and wonder.

The most concrete blurring techniques, which have evolved since the beginning of the last century, are collage and assemblage—bringing together disparate materials, forms, and images that already exist, rather than making them. In architecture, these can also be the building parts you find on site. These pieces are usually castoffs that have been liberated from their former uses—like the fragments of newspaper the artist Kurt Schwitters used in his collages, or the unused machinery populating the Zollverein, a former mine and steel mill turned into a cultural center in Essen, Germany—but that still bring some of the associations of their past function with them. In the collage, you then put the images together in a way governed by compositional principles, rather than functional hierarchies. You do so in a manner that implies that this assembly could continue far beyond the piece's picture frame. This approach has been by now perfected by artists. It is not yet common in architecture but is of particular relevance in the imaginative reuse of existing buildings.

Collage incorporates both multiplication and blurring, but it is also a form of active archaeology that collects remnants after they have been found. It is thus a modern version of the ancient act of hunting and gathering by which humans not only made ourselves at home in the world but also made sense of it. Collage presents an order that is rambling, unstated, continual, and open to interpretation. It does not tell you what is more important or what was produced when and how. It accepts and finds relations between things, reusing them in ways that allow for multiple interpretations. An unused wrench can show up in a collage as a purely formal element that still evokes work; old machinery like that at the Zollverein is impressive in and of itself, while also evoking the glory and grime of the Industrial Revolution.

TOWARD A CIRCULAR ECONOMY

This form of evocative collecting and building is particularly relevant as we realize that an important part of how we have to build today is by reusing not just buildings but their bits and pieces. It is not possible to reuse every building as it is. In those cases, we must understand building as part of the circular economy in which nothing goes to waste and everything is reused. Even though that is an ideal, it is remarkable how close we can already come: in the Netherlands, the architect Jan Jongert has created new buildings with over 90 percent reused and upcycled materials. Often the difficulty is not finding the actual materials but figuring out how to disassemble and reuse the old buildings. Jongert has figured out that worn-out windmill blades can make excellent seating and that discarded washing machines can be picked apart to build up new building fronts. In some cases, materials must be not only disassembled but also detoxified. This again is something we are learning in other parts of our economy—where waste streams also have to be cleaned up—and are now applying to (re)construction practices. Of course, that also means we have to put our newly remade structures together in ways that allow for easier disassembly for future instances of reuse. We might, for instance, assemble structures with visible connections that can be unbolted or unscrewed, rather than gluing or laminating surfaces together.

It is not only buildings that we can mine for used parts. All around us are heaps of things we have discarded, from cars to appliances to furnishings. One architect, Greg Lynn, in the early years of this century even developed a computer program to help him turn his children's old toys into building blocks for walls. When we use material that was not intended to be part of a building, we have to be clever about how to alter, fit, and adapt it. It is a question not just of recycling but of upcycling.

Instead of hunting and gathering in forests as our ancestors did, we now go wandering through cities and their outskirts for building materials. Instead of mining ores or extracting oil and gas, we engage in urban mining. There is a whole mini industry of designers and theoreticians at work figuring out how best to do this. Along the way, they are establishing best practices that future generations will be able to use. They are also advancing a new attitude about what architecture is, where it begins, and what it means, as the German architect Daniel Stockhammer notes: "The preservation and qualitative reuse and repurposing of existing building stock means architectural relevance is gained through complexity and multiplicity of meaning (instead of through form)." Not only that, Stockhammer says, but "the simplicity, durability, and sustainability of building construction, building materials, and technology are challenged and promoted." The process also redefines the role of the architect in a positive manner: "For the designers of our built environment, treating architecture as project (and the intellectual property) of many generations entails a transformation from creator to contributor." In that role, the architect must think not about abstractions but rather about the complexities of everyday life, asking, for instance, "How do we conceive designs if indeterminacy becomes an essential component of architecture? Which design approaches and tools are suitable for dealing with irregularities, incompleteness, and deficiencies?" The final question becomes not one of how long the building will last but rather: "What structures and materials are suitable for keeping the processes open to enable that future buildings and construction waste can (again) be reused and repurposed?"[10]

This urban mining is a kind of reverse potlatch. Instead of handing out new objects or food, as in those communal feasts held by First Nation groups in the Pacific Northwest to redistribute excess resources, urban mining finds and makes available what we have discarded. Part of how our capitalist economy works is by producing ever newer and more numerous products that lead us to throw away old things and accumulate more than many of us can actually use. Taking all that discarded excess and reusing it is a way of reversing this waste. Recycled material also brings with it the memory of what it originally was, similar to how the machinery in renovated factories evokes the Industrial Revolution. Good architecture—as opposed to just upcycling building practices—activates that knowledge. As the critic Andreas Hild has put it: "The city can be read as a reservoir, a stockpile of material also of sign. In the built substance, signs and matter interpenetrate each other to become an inextricable amalgam." The answer to this problem is urban mining, which, Hild says, "is therefore not merely a repurposing of material resources, but also requires a conscious approach to the world of existing signs and the memories bound within them."[11] When we put found objects together into built form, they should retain as much as feasible of what they were, while also becoming part of the world they are helping to make possible.

For the last few decades, critics and architects have been explaining these techniques using the analogy of the palimpsest. Strictly speaking, a palimpsest is a manuscript in which somebody has written over or between the original text, partially erasing or obscuring it, while letting enough of it remain that you can at least begin to decipher what was once there. The original Latin term referred to medieval manuscripts that people reused by erasing old text so they could save precious parchment; the term was brought back to life by deconstructivist critics such as Jacques Derrida in the latter half of the twentieth century. They used it to describe their works of criticism as a form of interpretative and evocative philosophy that did not state either itself or its ideas as new, but rather insinuated other ways of thinking within existing texts. Those texts could be literal books but also images, laws, or even buildings. In one famous example, Derrida riffs on a sewing machine that magically becomes the textile on which it works and also the text describing it. The architect Rodolfo Machado has claimed that all of architecture can be described in this manner. Whether or not that is the case, it is certainly true for almost all forms of imaginative reuse.

The palimpsest, Frankenstein's monster, blurring, and any of the other ways of describing imaginative reuse are all useful tools for helping us understand ways in which we can make architecture out of that act of renovation or reoccupation. In this book, I will show how such work is already happening in many different ways and all around the world.

REMNANTS OF THE INDUSTRIAL REVOLUTION

In contrast to the more traditional fields of historic preservation and adaptive reuse, imaginative reuse's core is not monuments such as churches or palaces but the grandest remains of the Industrial Revolution. Rediscovering the beauty of how we used to make things is at the heart of this form of regeneration. The grandeur of these factories, steel mills, and mines, as well as the expressive nature of their parts, offers something that stands outside of our daily lives in both scale and character. Just as we could gaze up in awe in naves of cathedrals, so we can stand in wonder in the remains of factories. Just as we can marvel at the intricacy of decorative schemes in the rooms where the wealthy once lived, so we can admire the gauges, levers, pipes, and rivets of boiler rooms. The remains of the industrial world directly ask us to consider the transformation of the Western economy from one based largely on the making of things to one based on the experiences we have. At the same time, these grand industrial workplaces were not originally meant to impress us or leave us in awe, which is why they awaken emotions that are closer to a fascination that mixes dread, admiration, and an interest in how things worked. These structures—open and expansive, with ample light and ways to get in, out, and around them—are also flexible and have fewer legal restrictions on how we can use them. At the same time, the intensity of the activities that used to take place in these industrial sites means that they have layers of patina (or simply grime) that makes that passing of time clear.

At a smaller scale, the flexibility of old industrial workplaces means that they turn out today to be the most easily adaptable spaces to live, work, and play. These old spaces are most commonly transformed into lofts, and the loft as a module can be turned into just about anything. Because the loft is so minimal in its design—often being no more than roughly built or poured walls, floors and ceiling, structural elements, and grids of windows or skylights—it fits well with lives that are turning away from a concentration on possessions and toward minimal objects with maximal uses.

From these core objects—the monuments and vernacular of the Industrial Revolution—a different kind of reuse has emerged, one that preserves the scale and texture not of palaces and churches but of factories. These old factories, warehouses, and workplaces now house cultural institutions such as museums and galleries, as well as libraries, community centers, and other spaces for social enjoyment. From there, it has been only a short jump to installing in these remains of our productive economy more commercial consumption sites as new stores, shopping malls, offices, restaurants, bars, cafés, and other sites where we shop or play.

THE ROLE OF ARTISTS

The imaginative part of this form of imaginative reuse had another root as well. As I will show, artists—from the collage makers of the 1920s to those who created art installations in the 1990s and beyond—have questioned the very act of making and meaning, and they have tried to understand how to place themselves in their world, reinterpret it, and open it up. In so doing, they began turning the practice of renovation into works of art. They saw reuse as a form of creative making. They were also the pioneers who first mined buildings for their expressive capacities: the stories of the lives that were lived there, and the roots and nature of the places that had generated them. Much of this work critiques our society and culture. It takes advantage of the fact that architecture is a social practice—not only in its making but also in its use and reception—to catalyze neighborhood change. Renovating disused buildings is one such example. Architects often like to think they can do the same thing when they design brand-new buildings, but the very newness of their products makes them appear like alien beings landing in a community.

Artists are the ones who have developed the scraping, enhancing, and aestheticizing practices that constitute much of imaginative reuse and that stand in contrast to the rougher forms of subtraction, addition, and elaboration, which architects have long practiced. They have incorporated cast-off materials into works of art and made us see them as beautiful (such as Joseph Cornell's compact accumulations of bits and pieces, including antiques and found fragments, placed in small boxes that resemble religious icons); they have assembled collections of record albums, magazines, and chairs and made them into art works, as Theaster Gates does; and they have, as many since Picasso have done, painted over, scraped away at, and added onto what they found on the street to turn it into art. The contemporary painter Mark Bradford is a recent example: he collects handbills from the street around his studio, pastes them on canvas, slathers them with automobile paint, scrapes away at the result, adds more found forms, then paints over them, then scrapes away to partially reveal them, until the many layers together form what he thinks of as a map of his native Los Angeles.

Following these examples, the artful form of renovation proceeds with respect for what is already there but does not leave it alone. It peels away layers, questions relationships, opens cracks and fissures, and turns things that were once useful into images of themselves—integrated paintings or sculptures. Artists have made such art in buildings that they have renovated. They also have turned whole buildings into artworks or, going beyond that, made installations out of the remains of buildings, created projections or fantastic evocations of buildings, and even created completely fake buildings. These artworks question not only what a building is and what its purpose might be but also its time period, place, and even reality.

Over time, these artists' experiments have become more widespread. Their work has seeped first into private homes and apartments, which are where architects have the greatest number of chances to try out ideas. You can now find houses built inside of ruins, preserved and stabilized, with new elements added to make sure that the owners are comfortable. You can find houses assembled out of upcycled building materials. You can even find concrete office buildings with irregular openings that let light and air into these formerly closed boxes so that they can be comfortably occupied. These inhabitations in ruins have been heavily influenced both by "high tech"—an art movement begun in the 1970s that delighted in exposed structure—and the nostalgia for past grandeur that emerged toward the end of the twentieth century, celebrated in lifestyle magazines such as *World of Interiors*. These styles became ways in which architects showed that you could inhabit the remains of the days, enveloping yourself in history and transforming it into elements of your daily rituals. I would argue that it has taken several decades for imaginative reuse to spread from the grand monuments of the Industrial Revolution (such as the renovated steel mills and coal mines in Ruhr, Germany, which began in the late 1990s) and intensely designed private homes (most notably 500 Capp Street in San Francisco, the house-cum-art project that David Ireland started creating in 1974) into our mainstream consciousness. By now, appreciating the beauty of the old—used and tarnished, composed and collected, revealed in all its layers, and turned to new uses—is commonplace. We have also begun to feel the realities of climate change in our lives, so that the pressing need to conserve nonrenewable resources has hit home. A growing awareness of the ways in which we fabricate histories and privileges for ourselves—whether in politics, economics, or architecture—has led us to wonder how we cannot just accept or reject our inheritance wholesale but radically alter it. Between the spread of beautiful examples of what is possible and the ever more pressing need to adapt our buildings in ways that open them up for both use and meaning, imaginative reuse has become more and more prevalent.

IMAGINATIVE REUSE AND SOCIAL JUSTICE

The divide between reused buildings and brand-new buildings is also a spatial one: more and more, the suburbs are where we build new, and the inner city is where we reuse what exists. That difference also highlights the ways in which reuse has become a class issue. Loft renovations and repurposed buildings—which are usually more expensive than new construction—are the province of the wealthy. We have to realize that in the United States any form of renovation is almost inevitably an act of gentrification.

The next frontier will be the reuse—I would hope in an imaginative way—of our suburbs and our exurbs. That process will confront a whole set of issues, as most buildings in those areas have been constructed in modes of mass assembly and were not designed to last very long. Reusing a strip mall, a distribution center, or a McMansion will be much more difficult than reclaiming a loft. Already, the largest monuments of the first age of consumerism—the suburban shopping malls—are being mined for imagery and practical uses, such as community colleges or offices. Some architects are producing so far only theoretical projects showing how we can break open and imaginatively reuse suburban homes: the designer Keith Krumweide, in his *Atlas of a Another America* (2017), proposed turning those McMansions into communities, or interconnected homes around shared communal space, simply by rearranging the interior elements and the homes' placement on building lots. I know of nothing that has been constructed to this day that I would consider a best practice, but we can only hope that such realizations are just around the corner.

Here, too, the challenge will be how to renovate in a way that avoids gentrification. I am not sure there is an answer for that problem in our current economic system. We produce or reuse buildings for those who can pay for them, and we have very few mechanisms indeed to make those structures available to people who cannot pay but still need them. What architecture can do in our current system is wring the most space and design out of the minimal resources we do invest in provisions like social housing.

Luckily, the discipline of architecture historically has a social consciousness baked into it. Most architects are trained to believe that they are not only there to serve clients but also to build a better world. Even if that pursuit is, for most of them, confined to evenings and weekends, or even just their guilty consciences, there are enough talented architects trying to figure out how to use techniques like urban mining, dumpster diving, and building reuse to give me hope. It is up to architects to use their skills and knowledge to not only make cheap housing but also transform it into beautiful, just, and sustainable spaces that can make all of us feel at home in our modern world.

The danger is, first and foremost, that we will not make the necessary commitment to reuse in a manner that opens each site both socially and in terms of its meaning. In other words, my fear is that we will not invest in making these renewed structures available to all. In that case, imaginative reuse will become a style. There is nothing wrong with styles in architecture or any other art practice. They allow a maker both to identify themselves through their output and also to work in a historic tradition that makes their products easier to accept and understand. What is less interesting to me is when a style becomes a coating that makes it appear as if the designer is doing something that they are not. The example of *faux* new lofts and industrial chic, both of which started to appear as early as the 1980s, show how quickly reuse can turn into a coating on newly made things. I hope and trust we can do better.

The examples I have collected in this book will show how imaginative reuse can be liberating and rooting. It can make us feel at home in the modern world and help us understand where we have come from, where we are, and where we might be going. It can do so in the spaces we inhabit every day, with structures that set the scene for our social and personal lives. It can construct a new world around us that is strangely familiar and open to many possibilities. I find the architecture I have tried to excavate and display here exhilarating, perhaps a bit weird, but certainly beautiful and full of promise.

PART I

FOUNDATIONS

CHAPTER I

THE ARCHITECTURE OF REMAINS

A Short History of Reuse

THE SPOLIA OF ANCIENT ROME

Reusing structures and materials is a practice as old as building itself. Nowhere is this more evident than in Rome—a place known, not coincidentally, as the Eternal City. Walk down any street there and you can see how eternity is constructed: by taking a piece of marble from a temple and using it either as the floor of a church or the wall of a noble mansion. In this city, Catholicism's central place of worship, you can walk down the aisles of some of the oldest and most beautiful churches—such as San Lorenzo Fuori le Mura of the sixth century and Santa Maria of the fourth century—and you will notice something strange: some of the columns, both along the nave and elsewhere, are of different colors and materials (you can see the different elements in image #1 in the photo insert). They are all the same height, but some were obviously not hewn from the same quarry as the building. Dig into their history a bit, and you will find that they come from two different places—the tomb of the Emperor Hadrian, also in Rome, and the temples where ancient Romans once worshipped their many gods. The columns were transported to these sites for use after the older structures fell into disrepair and as the growing Christian communities fueled a need for more churches. Scan the walls of either of these churches and many others in Rome, and you will find the remains of the Roman empire—from the bases of the columns to their ornamented tops, and from parts of friezes to simple stones—built into the very fabric of the site. The whole city is a collection of fragments that have been reshaped and rebuilt out of the ruins of the past. The Romans used the word "spolia" to refer to the stones they took from abandoned structures, or from buildings built by those they conquered: they were the spoils of war, but also the leftovers that could be reused to make new and grander structures. This process of taking materials from one building to construct another is perhaps the most straightforward form of architectural reuse. Look carefully at a curving street in Rome's historical center, such as the Via Teatro di Marcello, and you will notice that its buildings, as the name of the street implies, were once actually part of the amphitheater of Marcellus, dedicated by Emperor Augustus in 12 BCE. The structures of this old amphitheater help house apartments today, and their shapes still define the form of the whole neighborhood. You cannot throw a stone in that city without it either being a piece of spolia or hitting one.

Apart from using spolia, another way to put old or underutilized buildings to work is to reuse them whole and change their function. That sometimes takes some adaptation, as when the Castilians built a Gothic church inside the Grand Mosque of Cordoba when they conquered the city in 1236. The Hagia Sophia in Istanbul, one of the most astonishing domed structures in the world, was built as a Christian church in 537, then was turned into a mosque when the Ottomans took over in 1453. It became a public site in 1935 and is now a mosque again.

While the practice of reusing materials and whole buildings is old, the science of it has evolved dramatically. When people cast around for spolia or other building materials throughout history, they did so with simple pragmatism. Whatever was useful was used, and whatever was in the way was torn down or out. There was little need to preserve other than to save money or time, and the past and its building blocks were usually seen as a much as restrictions as they were resources.

LOFTS—SPOLIA BECOME ART

It was only in the nineteenth century that these forms of reuse became their own discrete art. As mass-produced modernization took over more and more of the world, a nostalgia for the old, the familiar, and the well-worn also took hold. But, more than that, there was also the need to make a place for oneself in the modern world in a way that made sense, was affordable, and transformed the mass produced and the new into the particular and the familiar. No building type exemplifies that better than the industrial loft.

In fact, you could say that the single most successful and ubiquitous bit of architecture in Western nations—and now much of the world—is not the single-family home, the skyscraper, or the shopfront, but the re-used loft. The first loft buildings were constructed by the millions at the beginning of the nineteenth century—the very dawn of the Industrial Revolution. They first appeared in and around Manchester, England, and some parts of France. Designed to facilitate manufacturing, they were big and strong enough to contain machinery, flexible enough to allow for changes in the production process and for the continual movement of goods and people through its spaces, and sunny and airy enough to maximize worker production with as few physical obstructions as possible.

By the 1930s, however, industrial production had evolved and began to demand a different type of workspace: larger, more expansive, with easy access to transportation. Industries transitioned from those first-generation lofts to large-span steel factories situated outside of congested cities. Starting in the 1960s, artists in places like New York began reusing the old lofts. They found that the well-lit spaces were large enough for their own forms of production, with freight elevators that could hold their paintings and sculptures. Moreover, because these lofts were now useless to industry, they were cheap. In our current moment, we are now reaching back to the older practice of reusing components, turning to spolia, which these days increasingly include recycled wood floors, steel beams, and even carpet tiles.

I remember my uncle, who worked in a printing plant in Manhattan's Soho neighborhood, telling me in the early 1970s that some messy painters had moved into the floor above his business. A new bar had also opened down the street, and the character of the area was changing. Soon my uncle's firm had decamped for New Jersey and the artists had taken over. In the 1980s they were replaced by offices for lawyers, accountants, businesspeople, and regular wage earners, and then, in part, by Google, which, after working in various buildings around Manhattan, consolidated its New York outpost in 2022 in a Chelsea building where books were once produced.

The loft is a perfect example of the reality of reuse. It is flexible by its very constitution, and there are a lot of them, which also means the forms are standard and familiar. This allows there to be clear rules and mass-produced ways to adapt the lofts; a whole industry has grown up to feed the continual renovation of these former industrial spaces into places to live, work, or play. Lofts developed a certain aura of creativity, exactly because artists were involved, and then the places where they showed their work (galleries) and where they played (bars, nightclubs) pioneered the movement of reusing them. We think of lofts as slightly artistic and avant-garde, even when they have been completely gentrified into sophisticated residences for wealthy inhabitants.

There is a larger story here as well: the loft marks the transformation of our society from one of production into one of consumption—a flip first noted by critics and theoreticians of social movement, such as C. Wright Mills and John Kenneth Galbraith, even before the first residential lofts came into use. These buildings were once factories and large workshops, the productive engine that modernized the world and lifted into the middle-class over 50 percent of those in the West and, more recently, a third of those in Asia.[1] The contemporary loft—either as creative workshop, apartment, or shopping mall—is the emblem of a new economy, which is based on the consumption of products that are now manufactured in anonymous warehouses somewhere far away, either in the countryside or in other nations.

If spolia and reuse have always been part of making buildings, what lofts brought with them—even if they did not use spolia beyond leftover mechanical equipment and columns—was the acceptance of the original form and the aura, converted into a stage set for daily life. With the loft, reuse became a matter of valuing the past in a particular way—or rather, in many ways that, taken together, have energized the newly built environment with the energy of the old.

HISTORIC PRESERVATION AS NATION BUILDING

The larger idea of historic preservation—restoring historic buildings in order to preserve them as they were (or might have been when they were first built)—has its roots in Europe and the United States of the late eighteenth century. The trend was both an expression of scientific and social progress and a defense against the gathering soot clouds, functional factories, and housing estates of the Industrial Revolution. It was both a reactionary movement, trying to preserve or recreate what was under threat from modernization, and a revolutionary one, attempting to create a scientifically accurate and stable evocation of another time in a way that could activate current uses.

In France, those two impulses became intertwined, especially in the first intendant of historic preservation, the architect Eugene Viollet-le-Duc, who lived and worked in the nineteenth century.

Viollet-le-Duc was clear about his purpose and its importance. "Our age," he wrote,

> has adopted an attitude towards the past in which it stands quite alone among historical ages. It has undertaken to analyze the past, to compare and classify its phenomena, and to construct its veritable history, by following step by step the march, the progress, the successive phases of humanity. . . . Our age is not satisfied with casting a scrutinizing glance behind it; the work of retrospection cannot fail to develop the problems presented by the future and to facilitate their solution.[2]

The act of restoring was one of interpretation and resurrection, but also modernization:

> In restorations there is an essential condition that must always be kept in mind. It is that every portion removed should be replaced with better materials, and in a stronger and more perfect way. As a result of the operation to which it has been subjected, the restored edifice should have a renewed ease of existence, longer than that which has already elapsed.[3]

Originally a firebrand revolutionary, Viollet-le-Duc forged his greatest legacy in the restoration of three medieval structures: Notre Dame in Paris (restored between 1844 and 1864), the royal castle of Pierrefonds (1857–1860), and the walls of Carcassonne (1849–1859). Of these, Notre Dame is the most famous, especially given the recent controversy over its restoration after the 2019 fire. Immediately after that fire, which destroyed the roof and spire over the main nave and severely damaged much of the interior, architects began pushing for an international competition to find a design that would give contemporary architects a chance to reimagine it. They were interested especially in new visions for the spire, taking into consideration current technology and ideas about religion (Notre

Dame is, after all, France's central cathedral) and national spirit. Public reaction quickly squashed the idea, and the country's parliament instead mandated that the spire should be rebuilt exactly as it had been. The irony here is that Viollet-le-Duc had invented that feature in 1844 out of whole cloth, using what historic documents he could find and studying other Gothic structures to design what he felt was a correct punctuation to the national shrine.

Viollet-le-Duc supervised the restoration of the cathedral as a symbol of renewed French might and unity under Napoleon III. He used what technology was available to him but tried to make what he did appear as an organic part of the overall building. It was as much an attempt to evoke the Notre Dame of Victor Hugo's (still) immensely influential 1831 novel as it was a work of necessary renovation to make the cathedral more stable and weatherproof. As he said of his work there:

> It is necessary not only that the artist apply himself to propping up, strengthening, and conserving; he must also make every effort to restore to the building through prudent repairs the richness and brightness of which it has been robbed. It is thereby that he can conserve for posterity the unified appearance and the interesting details in the monument that has been entrusted to him.[4]

The political and cultural project in Viollet-le-Duc's other major restorations was even more overt. At Carcassonne, an enclosed hilltop city that had once been one of Europe's impregnable fortresses, he resurrected the walls that had historically surrounded it. He did so again on scant evidence, extrapolating from the existing ruins and creating what is as much a fairytale vision of a medieval town as it is a scientific reconstruction.

When Viollet-le-Duc transformed the ruins of the former Pierrefonds royal hunting lodge into a palace for Napoleon III, the effort was a purely political one. Napoleon III, who, like his uncle the emperor Napoleon, had taken power initially as a revolutionary and then had himself made king, needed to establish his legitimacy as the royal ruler of a fast-industrializing France that had grown conflicted about the whole notion of kings and queens. What Viollet-le-Duc did was to connect Napoleon III to the older regime by grounding his collection of Louis XIV furniture and medieval tapestry in the former hunting lodge and royal residence.

To do that, he had to completely reinvent the building. The original castle had been constructed in the fifteenth century, destroyed by Louis XIII, and was nothing but a ruin when Napoleon III found it. Viollet-le-Duc reinvented it as a fairy tale site, bristling with turrets, tall walls, a moat, and a drawbridge. He deliberately designed the whole to look as if it had been built over the ages, with additions and juxtapositions of various medieval and early Renaissance styles. Only the uniformity of the stone gave away the fact that it was mostly a new structure.

Around the same time that Viollet-le-Duc was doing his reconstructions, the Dutch nation was building a royal museum and train stations as part of a conscious effort to both remake itself as a modern nation and recall its so-called Golden Age—the period from the mid-sixteenth to the mid-seventeenth century when it had been the most powerful economic and naval power in Europe. That age had produced the seamless facades of Amsterdam's Canal Area, which displayed the wealth of the merchant class in its continually curving streets that gave the illusion of a complete world, opening to vistas beyond the next curve and bringing the viewer back to the continual faces of the rows of houses. Most of the zone was in ruins or had been built over by the middle of the nineteenth century. One family of three generations of architects, the Koks, starting at the end of the nineteenth century, rebuilt hundreds of the houses according to what was left of historical records, the meticulous paintings of seventeenth-century artists, and their own conjecture. When the millions of tourists who now traipse through Amsterdam every year admire the beauty of its merchant world, they are actually seeing an idealized reconstruction of what might have once been there.

Reconstruction and historic preservation became a specialty within the practice of reusing buildings and their parts. And that activity came with a particular agenda: to house the present—with all its newness and continual reinvention, all its technology and comfort—quite literally within the old. This reinvented historical architecture became the anchor of new states, of new institutions such as national museums and theaters, and of the now-dominant middle class, who had grabbed power in the parliaments and other ruling bodies of the West by the end of the nineteenth century. It made the modern world seem familiar, possessed of authority and the aura of a history reaching back beyond a recent past in which many of these countries had been ruled by foreign powers.

PRESERVATION VERSUS RESTORATION

The thinker who contributed most to the notion of preservation as nation or culture building was the English critic John Ruskin. In his treatise on architecture, *The Seven Lamps of Architecture* (1849), he wrote:

> We may live without [architecture], and worship without her, but we cannot remember without her. How cold is all history, how lifeless all imagery, compared to that which the living nation writes, and the uncorrupted marble bears!—how many pages of doubtful record might we not often spare, for a few stones left one upon another! . . . There are but two strong conquerors of the forgetfulness of men, Poetry and Architecture; and the latter in some sort includes the former, and is mightier in its reality; it is well to have, not only what men have thought and felt, but what their hands have handled, and their strength wrought, and their eyes beheld, all the days of their life.[5]

For Ruskin, in contrast to Viollet-le-Duc, that meant not restoring but preserving, thus setting up a debate that continues to this day:

> Restoration . . . means the most total destruction which a building can suffer: a destruction out of which no remnants can be gathered: a destruction accompanied with false description of the thing destroyed. Do not let us deceive ourselves in this important matter; it is impossible, as impossible as to raise the dead, to restore anything that has ever been great or beautiful in architecture. That which I have insisted upon as the life of the whole, that spirit which is given only by the hand and eye of the workman, never can be recalled.[6]

For Ruskin and his followers, this idea of preservation extended into making new buildings that sought to preserve that spirit. They believed that only neo-Gothic architecture—even if carried out in new materials and housing new forms—could resist what they saw as the soulless and placeness nature of the modern city and its tentacles reaching out into the countryside.

Some of Ruskin's followers were less interested in making new buildings that continued medieval techniques, focusing instead on preservation. In 1877 the designers William Morris and Philip Webb—leaders of what came to be known as the Arts and Crafts movement, which sought to reimbue the making of everyday objects with meaning—published their *Manifesto of the Society for the Preservation of Ancient Buildings*, in which they wrote:

> It is for all these buildings, therefore, of all times and styles, that we plead, and call upon those who have to deal with them, to put Protection in the place of Restoration, to stave off decay by daily care, to prop a perilous wall or mend a leaky roof by such means as are obviously meant for support or covering, and show no pretense of other art, and otherwise to resist all tampering with either the fabric or ornament of the building as it stands; if it has become inconvenient for its present use, to raise another building rather than alter or enlarge the old one; in fine to treat our ancient buildings as monuments of a bygone art, created by bygone manners, that modern art cannot meddle with without destroying.[7]

The core of their work was thus the preservation of buildings that they felt were important. We usually define such structures as monuments, a general term most clearly defined by the German theoretician and historian Alois Riegl in 1903 in his "The Modern Cult of Monuments; Its Character and its Origins."[8] Riegl wrote that a monument was "in its oldest and most original sense . . . a human creation, erected for the specific purposed of keeping single human deeds or events (or a combination thereof) alive in the minds of future generations."[9] Any one of three conditions could make something a monument: a significant event that had occurred there, including the life of an important person; something that had been built to memorialize an event, structure, or government; and something that had survived long enough to, by its very endurance, become a monument.

THE VENICE CHARTER

Viollet-le-Duc's campaign of restoration, Ruskin's love of ruins and remains, and Riegl's definition of monuments composed much of the background for the often-intertwined practices of restoration, preservation, and adaptive reuse that arose staring in the middle of the nineteenth century. The definitions of how to engage in restoration and preservation were codified in 1964 in a document that has come to be known as the

Venice Charter, which brought together experts in the field under the aegis of the United Nations.

The Venice Charter breathes the spirit of Ruskin:

> Imbued with a message from the past, the historic monuments of generations of people remain to the present day as living witnesses of their age-old traditions. People are becoming more and more conscious of the unity of human values and regard ancient monuments as a common heritage. The common responsibility to safeguard them for future generations is recognized. It is our duty to hand them on in the full richness of their authenticity.[10]

It also allowed, however, that:

> The conservation of monuments is always facilitated by making use of them for some socially useful purpose. Such use is therefore desirable but it must not change the lay-out or decoration of the building. It is within these limits only that modifications demanded by a change of function should be envisaged and may be permitted . . . Replacements of missing parts must integrate harmoniously with the whole, but at the same time must be distinguishable from the original so that restoration does not falsify the artistic or historic evidence.[11]

These standards were adapted by most countries around the world, including the United States, and now form the background for the reuse of most buildings or sites we consider significant.

This history of the preservation and restoration movement also makes clear that these efforts were integral to the way modern societies sought to define themselves by housing themselves in the past. Nowhere was that truer than in the United States. The nation's very newness, as well as its economic and soon political success, made it necessary for it to find a way to root itself. This was especially true after the Civil War, which had almost destroyed a nation then less than a century old. To trace the ways in which the country reimagined itself, no bit of architecture is more appropriate than "the nation's home": the combination home, office building, ceremonial party location, and icon called the White House.

THE ARCHITECTURAL MYTHOLOGY OF THE WHITE HOUSE

A picture from 1952 sums up the type of renovation that the White House underwent over and over again. The photo takes us into the empty shell of a building. We see the outside walls of the White House stripped of the corridors of power—there is no Oval Office, no presidential bedrooms, no gathering sites of the high and mighty. All that you see are outlines of the building—making you realize how big it really is—held up by steel struts. The photograph is just one point in a continual remaking of what we think of as an unchanging national icon—a remaking that has been ongoing since a few years after it was first constructed in 1799. Only the first rebuilding was occasioned by destruction: the reconstruction after the British burned down the brand-new White House in the War of 1812. Ever since then, each building intervention has been undertaken to fix structural problems, add amenities and facilities, or, more often than not, make the interiors conform to a president's vision of an exemplary home. Above all, every redecoration, reconstruction, and renovation has rebuilt the image of true America at a domestic scale, using the bones of what had originally been a relatively modest presidential home base.

Starting with the Grant administration in the 1870s, presidents and their spouses have used the White House to propose how the most powerful family in the country should live. For Julia Dent Grant and Ulysses S. Grant that meant living like New York robber barons of the Gilded Age: opulent fabrics, overscale furniture, and a sense that every aspect of every room should be grand and luxurious. In 1882 Nell and Chester Arthur brought in the multifaceted Louis C. Tiffany to give the place a more coherent and stylish interior that also displayed the talents of a new generation of American designers.

It was under Theodore Roosevelt, president from 1901 to 1909, however, that the ritual of White House redecoration turned backward toward reconstructing an imagined past. Roosevelt's administration came at the end of what the historian T. J. Jackson Lears has called "the antimodern impulse."[12] The traditional American elite, of which Roosevelt was very much a part, felt threatened by the ascendancy of new money (as opposed to the inherited or slowly accumulated wealth of the East Coast) held not only by those in "the West" (the Midwest) but by people of nonwhite ethnicities and origins. This traditional elite also believed that the means by which newcomers were making their money—from factories to railroads to department stores—was ruining the nation's landscape.

Roosevelt's response was to resurrect what he saw as America by founding the first national parks: restricting the robber barons and breaking up their cartels; exhibiting and arguing for the "manly" virtues of exercise, camping, and war; and supporting artists, craftspeople, and designers who were trying to reconstruct a pre-Revolutionary architecture and décor. The White House—which he and his wife, Edith, had redecorated by the New York firm McKim Mead and White—was the emblem of these efforts.

Under the Roosevelts, the style of the White House became, and has been ever since, neo-Federal: a reimagination of what the building might have looked like when it was first rebuilt by the Adams after the War of 1812, with touches that either reached back to a pre-Revolutionary era (under the Kennedys, for instance) or added grander touches that actually had an aura of the Gilded Age, as with both President Nixon's and Trump's decorators. Whatever the tendencies, the message was the same: American history was defined by the wealthy planter and trader elite of the eighteenth century, and we are just living in—and redecorating—the house they built.

The White House renovations are most visible and are rhetorical statements of what has long been the central driver of historic preservation: to keep, restore, and turn into icons what is left of an imagined past as seen by those with the money and power to do the restoration.

In other contexts, however, historic preservation has been a revolutionary act that transforms the seat of one bygone power into the home of the new guard. One of the signal acts of the French Revolution, in 1796, was to take over the royal palace of Versailles and turn it into what became a prototype for, and which eventually became, an actual public museum. Similarly, today the plantations of the American South and even President Jefferson's home at Monticello have been renovated (often for the umpteenth time) to exhibit the formerly erased lives of the enslaved people who had labored there—and to acknowledge their forced contributions to the graceful structures of the Antebellum South.

RECONSTRUCTION AND THE REMAKING OF HISTORY

By the 1930s, the movement to renovate the structures that had come to be regarded as icons of the past turned into both a science and a political tool. Industrialist and philanthropist John D. Rockefeller paid for the meticulous reconstruction of what was left of Colonial-era Williamsburg, turning it into a showcase for America. The organization he helped fund hired experts to reconstruct what the town might have been, then paid actors to reenact the lives of townspeople. Even the food preparation and the crafts were reconstructed. Williamsburg eventually inspired the construction of entirely new towns, such as Disney's Celebration, Florida, (starting in 1996) and Seaside, Florida (from 1985 onward), that claimed to bring back traditional small-town America. Their elements—from front porches to small, right-angle streets rather than cul-de-sacs, as well as the extensive use of brick and wood—were architecture's equivalent of Ronald Reagan's campaign commercial promising a "morning in America," resurrecting not only small towns but their middle-class white values.

In Europe, some cities went even further. After successive waves of German and Allied bombings destroyed over 90 percent of Warsaw during World War II, Poland's new communist regime rebuilt the downtown area in concrete and steel—then made it look exactly like what had been there before the war. New functions and mechanical systems hid behind the facades of medieval and Renaissance buildings, and the years of German occupation were wiped away from public view. The communist regime in this manner made the case that it was not only building a more ideal world but also was doing it in a manner that resurrected the past glories of the Polish state.

Nor were the communists the only ones to try to ground their new state in the familiar and glorious forms of the past. On a smaller scale, the centers of Frankfurt in West Germany and Dresden in East Germany were rebuilt in a similar manner. More progressive local governments, such as those in Berlin, meanwhile, preserved and highlighted the remains of significant structures to keep alive memories both of the war's violence and of the evil regime that had necessitated it. In 1963, Berlin's Kaiser Wilhelm Memorial Church, bombed during the last days of the war, became an open and empty site elaborated with evocative grids of concrete and stained glass, constructed to a design by architect Egon Heinemann and commissioned by the city government.

Yet even while such ruins were being mined for their relevance, other grand edifices were losing the battle with modernity. Rotterdam, the Netherlands, which had also lost its city center to the violence of World

War II, was rebuilt in the decade after as a modern district of shopping, living, and working. This area followed principles of efficiency rather than preservation. In both the US and Europe, whole neighborhoods were wiped out to make way for expressways, new housing, and office developments. By 1976 in Albany, New York, a preserved state capitol looked out over rows of government offices in high-rises that flanked a reflecting pool, underneath which a massive parking garage sucked in a highway offramp. Beneath all of that lay the rubble of what had once been a thriving neighborhood, inhabited by working-class people of various races. In New York City, the Puerto Rican neighborhood made famous by the Broadway play and then the movie *West Side Story* was wiped out in 1959 for the construction of Lincoln Center. The original Penn Station, a vast edifice based on Roman baths, made way for a replacement composed of dense underground corridors, waiting areas, and train tracks.

A NEW PRESERVATION MOVEMENT

In 1975, when Penn Station's equally grand twin, Grand Central Station, was threatened with demolition (or at least major changes to its spaces and appearance), a movement was finally galvanized (at least in the United States) to preserve historic buildings—even if presidents hadn't slept there, even if the building didn't represent a colonial past. Led in part by the same Jackie Kennedy who had helped create a "purer" version of the White House when her husband was president, the movement spawned laws establishing what now is a powerful regulatory system.

In this country and throughout much of Europe, any building over fifty years old is now eligible to be preserved. The process for nominating and protecting such a building differs by locale, but in most locations it is difficult to tear down almost anything that has even an inkling of associated social value or a few columns and pediments to denote architecture. How it is reused is defined by rules set by the Venice Charter. The past is no longer a foreign country. It is a current reality in which we must all live.

Although these different attitudes succeeded each other, they have also existed together and persist today. The past is indeed a contested place, and how or what we preserve is a proxy for political, economic, and cultural beliefs. Although these beliefs are most evident in historic preservation, they have also seeped into other, less charged aspects of reuse: the adaptive reuse of existing buildings. That common practice has veered between preserving as much of an old building as possible—both for practical reasons and because a more antique structure offers a certain authority—and seeing the building simply as a usable shell, with little respect or regard for its history.

THE POLITICS OF REUSE

For some activists, the past is also a place to be conquered or at least squatted. During the 1960s, while governments were busy defining what a monument was, how buildings should be reused, and what all that meant for their country's self-definition, citizens began taking over historical structures on their own. The squatting movement, which originated in northern Europe at the end of the 1960s, was primarily the result of a severe housing crisis, especially affecting those with less means, the young, and the unemployed. Governments and developers couldn't build fast enough to meet the demographic explosion of the period, which was also the result of large-scale immigration into these countries. Meanwhile, countless existing buildings sat empty, either because they had been damaged by the war or because owners thought they could tear them down at some point and put them to better use. Squatting in them became an illegal and later semi-legal action that spread throughout the world. While squatting had little regard for the niceties of preservation, it tended to leave what was there alone, only seeking to make it more habitable. It also opened up buildings: squatters took down doors and gates, connected spaces for communal living, and in general cut into the grandeur and formality many of these older building possessed.

The squatters' attitudes toward reuse eventually became an integral part of the larger practice of reusing buildings. It pioneered ways to adapt structures not by demolishing but by small actions: subtractions, additions, cuttings, and elaborations.

HISTORIC PRESERVATION

Over the course of a hundred years, then, preservation evolved in different ways. Originally, it was a practice in which rubble was reused because it was there, or in which certain old buildings were utilized because they made sense for new uses. In the nineteenth century, preservation became a movement in which certain structures were preserved, rebuilt, or reimagined because of the role they could play in defining families, states, or other institutions. After the middle of the nineteenth century, reuse became an attempt to present a different future, and by the end of that century it was even considered a reactionary attempt to preserve a status quo fixed in inhabited form.

The techniques developed for historic preservation and adaptation were designed to either affirm existing power structures or to invest new ones with authority; however, they can also be used, as the squatters did, to question them. This means not just making the former spaces of the rich and powerful available to other groups, but also engaging in a form of redesign and adaptation that reveals and even counters the power structures that are so evident in them. This might mean making new openings in unexpected places or changing the scale of existing rooms. It might also involve revealing layers of history in ways that deny the pompous orders you might normally see on display in pediments, columns, and friezes.

In rural Virginia, one of the plantation houses once built and owned by the Lee family—which historically included not only Robert E. Lee but also the richest and most powerful slaveowners in the Commonwealth—is now being turned into a visitor center. The renovation adds glass and a few new spaces to what is essentially a ruin. The fragments of the past, woven together with elements that both stabilize and reveal them, become ways that we can understand the pretentions and the wealth of this family, and actively deny that power by re-inhabiting what is left of their structures.

In South Africa, the jail that once held the apartheid regime's political prisoners before 1994 is now the nation's High Court. During the period of the Truth and Reconciliation Commission, which sought to reckon with the crimes of the apartheid government and the violence that came out of opposing it, the court became a site where countless victims and perpetrators came together not just to tell their stories but also to rediscover places of violence and incarceration, which have since been transformed into deliberative spaces of knowledge and justice.

So what does "reuse" mean today? Although adaptive reuse, renovation, restoration, and reimagination all started from the same idea that we can and should preserve or reuse existing buildings, each process has become a distinct part of the larger movement to recycle, upcycle, and reuse the structures built by previous generations. Each of these approaches means something different for our society. Straight renovation has few rules and broadly means taking whatever building comes your way and making it work for your needs. Historic restoration and preservation, in contrast, is a field with a science of paint chip forensics and historical records analysis that allow architects to supposedly recreate a building from ruins or oft-renovated layers of occupation. There are gradations between these two poles, each defined by a web of local and national regulations and laws. But beyond these two traditions we are also now reaching back to the older practice of reuse of components.

The need to reuse what we already have is now imbedded in the way we think about almost every aspect of our lives. We reuse plastic bottles and paper bags, and recycling is part of our daily rituals. We also value older clothing and objects, often more than new ones—even when, as at stores such as Restoration Hardware, their age is faked. Many of the most striking new bits of architecture to appeal to the public imagination—from New York's High Line to Ponce City Market in Atlanta, from the 2022 Beijing Winter Olympics stadium to the new Bourse de Commerce museum in Paris—reuse humble, often industrial structures.

In Europe, developers are finding that older structures have become gold mines out of which the new apartments, museums, and offices can be minted.

At the same time, we are finding that we can tell new stories and affirm new identities by reusing and redesigning the sites of power we have inherited from our forebears. By revealing the hidden truths in the plantations of the American South and the palaces of Europe, they become scenes in which people can live their own lives. We are reimagining our past, present, and future through reuse.

CHAPTER 2

DUMPSTER DIVING AND URBAN MINING

The Materials of Modern Reuse

The Dutch architect Jan Jongert takes reuse seriously. As he walks you through the buildings he has designed, he tells you what percentage of the structure was made from materials repurposed from other building sites or even other uses. Usually, the proportion is at least 66 percent and often close to 90 percent. He has used discarded washing machine parts to create food stalls, trucks tires to make seating in a music club, and wind turbine blades to make seating and climbing sculptures in playgrounds. Jongert, an elfin presence darting through his projects, is the ultimate dumpster diver in architecture.

While his reuse percentages are impressive, Jongert points out that he is just following the guidelines for reuse that the European Union had set way back in 2008: 70 percent of all buildings should be constructed out of reused or recycled material by 2020.[1] Every country in the Union has missed those marks by a significant amount, so Jongert just figures he is one of the few law- (or at least guideline-) abiding citizens in the field. Dutch rules already demand that half of all building materials be at least made from renewable and non-extracted material, which extends the definition of reuse a bit, but Jongert believes that we can and must do better.

THE CHALLENGES OF REUSING CONSTRUCTION MATERIALS

The concept of circular use is not particular to the construction industry, but it is of particular relevance there exactly because it is so difficult. All of us can and do easily reuse everything from paper to metal, and we have gotten progressively better at separating our waste streams, developing specific techniques for different types of refuse that used to wind up in landfills. We also reuse clothing, though often in a way that keeps colonialism alive—for instance, by sending bales of it to developing African nations to be picked through, worn, or otherwise recycled by a modern version of ragpickers. In most countries you cannot legally dispose of toxic materials such as paint or oil anymore; instead, these are reused for everything from new lubricants to heating. But the stuff that goes into buildings—from steel beams to lay-in ceiling tiles and bathroom fixtures—is more difficult to recycle.

In the case of renovations, the problem starts with the disassembly itself. It is easiest to send in the bulldozer or set off the dynamite and then pile the resulting rubble into trucks on their way to a landfill. Taking apart buildings piece by piece so that the materials can be reused is a much more complex task, demanding more time and precision and presenting a logistical puzzle. Engineers are often wary of pre-used steel, for instance, as they do not know what stresses the beams or posts were subjected to previously. Reused wood may be warped, and each piece might have a different ability to carry loads. Connecting materials are another challenge. Nuts and bolts are relatively easy to unscrew, although they have often been painted over, but how do you remove layers of coverings that have been applied to wall or floor surfaces? Or, to point out a pesky problem even a casual DIY-er will have encountered, what do you do with all the nails embedded in wood?

The construction industry is developing techniques to take on each of these issues, although they are often still not very established. The next step is to figure out what to do with toxic materials. Asbestos and lead pipes are famous examples, but many of the finishes put into buildings from the 1930s until fairly recently were based on petroleum or minerals; if they were not already dangerous to handle when they were used, they either might have become so over time or would no longer meet safety standards today.

REUSABLE BUILDING PIECES AND STYLE

There is also a question of style. Most architects design structures under the assumption that the materials they will use are neutral in color, sized according to industry standards, and pretested for durability and appropriate use. Clients who order up a new building also believe they will get a building made according to prevalent norms, tastes, and safety standards. That often means sleek, abstract, and, again, neutral in form and color. Such a look is the opposite of how most recycled materials appear. Truck tires look like truck tires, and washing machines look like washing machines. You can cover all the pieces up with paint or panels, but then you are missing the point. A building made out of other buildings or cast-off objects is by its very nature a variegated assembly. Until we accept that architecture can look as varied and complex as the rest of our world, can weather and change over time, and will need to be repaired regularly—as is the case, for instance, with a car you buy used—we will struggle to sell buildings that are part of a circular economy.

Accepting the necessity of circularity also means coming to terms with the reality of economic costs. Until we price new materials to truly reflect their full cost—including what obtaining them does to both our planet's health and our own, and how their extraction relies not only on exploitative labor practices but also on built-in subsidies for everything from roads and railroads to energy—it will usually be cheaper to make a building out of unused, standardized supplies.

MAPPING URBAN MINING

What Jongert and the growing cohort of designers like him are seeking to do is to make the argument for such architecture by building it well and beautifully. They are turning dumpster diving into an art form that is as efficient and cost-effective as possible. They have turned to the Internet to help them do so: in 2012, Jongert and his then partners developed a piece of software they called a Harvest Map. It regularized a practice they had been perfecting in which they searched near their work area in ever widening circles for material they thought they could reuse. They surveyed junkyards and shops specializing in various kinds of recycled materials or parts, but they also scouted out places with large reservoirs of things like truck tires, which were just waiting to be buried in a landfill. They noted the nature of each piece or collection that they found and then made spreadsheets out of all the data they could assemble on those items' availabilities and costs. The map graded these potential building materials according to quantity and potential ease of reuse, but also by proximity to the user, as transportation can add quite a carbon load to any element.

The Harvest Map was designed to be free and fully interactive, although it is now difficult to access since Jongert's old firm sold it in 2019 to a third party, New Horizons, a Dutch "urban mining" company that sources and sells recycled building materials. However, others have developed similar systems. Several years ago, ROTOR, a Belgian collaborative group, started a recycling supply company called ROTORDC, which stands for ROTOR Deconstruction. This outfit has used its experience in reuse to "develop deconstruction techniques, logistical systems and remanufacturing installations for contemporary building materials, with a focus on finishing materials."[2] They have made a specialty of mining buildings that were originally built in the last few decades of the twentieth century, finding ways to reuse the lay-in ceiling tiles, plastic laminates, lighting fixtures, and even sanitary equipment that they unearth there. To buy what they have found, you can go to their warehouse in Brussels, but you can also log onto their website, where the pieces are presented in a manner that appears to deliberately recall the look and feel of an IKEA catalog.

In the meantime, digital media have advanced to the point that anybody can make their own site-specific Harvest Map or catalog without too much difficulty: the needed tools—a combination of sophisticated Boolean searching, mapping software, and analysis of carbon footprints—are increasingly available. Moreover, social media has become the most flexible marketplace available, letting buyers, sellers, and scavengers connect in real time on myriad platforms.

What architects such as Jongert are concentrating on now is not devising systems—which is the province of the new economy of sustainability experts with whom he often collaborates—but on designing in a way that shows how beautiful the results of recycling can be. With a team of about a dozen employees, Jongert is extending his reach from small-scale projects such as clubs and single-family homes to larger buildings.

INSIDE BLUECITY

Jongert's Rotterdam office is an example of such a place. Its site is remarkable in itself: it is tucked away into the lower floor of the old Tropicana, a rambling white building that opened in 1988 as an urban resort, sporting a luxurious swim club and sauna. The building stretches out along the edge of the Meuse River and runs along the dike in a difficult-to-grasp collection of curves, domes, and angled structures that dip down to the level of the river and rise up to taller spaces. Over time, the Tropicana added restaurants and even a disco, but none of it really worked. The enterprise, after changing hands and formulas a few times, went bankrupt in 2010. With its strange form and very particular spaces—what can you do with an empty swimming pool and fake beach?—it sat vacant for many years before Jongert persuaded an enterprising developer to let him try to fix it up and help fill its snaking spaces. It is now called BlueCity, and it has become, according to its website, a "circular economy hub" that leases spaces in the building at very low rates to companies that focus on reuse in various industries, transforming waste material into everything from beer to paper.[3]

Walk past the doors of BlueCity and you are confronted with an assembly of windows, doors, and pieces of walls that together define the various spaces where different companies are housed. There are also fragments of furniture and some bits and pieces within those that seem undefinable at first, but they all come together and reveal themselves to be new offices, workshops, and laboratories. Superuse Studios takes up one corner of the building—a prime location looking out over the Maas River (although almost all the spaces have a similar view). In another area is a brewer making their product largely out of recycled food waste, and in another is a restaurant whose menu also makes good use of scraps, leftovers, and gleanings. One BlueCity workshop is experimenting with making building materials out of mycelium (mushrooms), while through the recycled windows of another space you can see people trying to formulate paper out of waste.

The atmosphere is funky and rambling, and you will not find many straight angles anywhere. Instead, your eye keeps finding new compositions, whether it is the way the various pieces of lumber—sourced from different places—now fit together with the precision of stonemasonry, or the hopscotch rhythm of the collection of recycled window and door frames that now make up the wall of a row of offices. Each of these reused pieces also has its own texture, from the worn wood to the rusted steel of structural elements to the sleek wood panels originally intended for a new school but ordered in the wrong size.

This worn and fragmentary look not only creates its own attraction, inviting you to explore the interior constructions with your fingers as well as your eyes, but it's also a fitting framework for most of the businesses that use the space: the folks who work here form a crowd that usually eschews fashion labels and trendy styles. The building, meanwhile, brings its own layer of history and aesthetics into play. The original white-painted columns and beams, though worn and peeling, are still there, as are the curved glass walls and arched skylights. The stone borders that originally led bathers along tiled paths from lounge to swimming pool to dressing room are also still in place. Even some of the tropical plants and the paintings of jungle scenes meant to take visitors far from gloomy Netherlands are still visible. Finally, there are the new mechanical systems, the cuts through some of the walls, and the bits and pieces of newer furniture that make the time period of BlueCity even more uncertain. The whole is a mess, but an endlessly fascinating one.

A good example of Superuse Studio's work is the Country Seat Brienenoord. Situated on a small island in the middle of the Maas, the site was originally a camp where children from working families could come to enjoy the great outdoors. It had been condemned for use when the organization that owned it came to Jongert. Superuse Studio renovated the building into a youth center, reusing 90 percent of the original structure. Everything they added—except for five trusses, a few fasteners, and bases for columns—was recycled. You would not know it from the way the place looks now. The building consists of interlocking trapezoidal spaces that rise toward the island's central open area. Clad partially in a translucent corrugated material, it only gives you a hint of what goes on inside. As the roof lifts from the ground in diagonal leaps, it makes way for stacks of reused windows in white frames, and through those you can get a sense of the space within.

The central room is a place of gathering, its columns supporting trusses that in turn help the angled forms of the roof reach their crescendo. Within the wood and steel armature, glass, and corrugated walls, kids are playing, helping to prepare food in the open kitchen, learning about nature, and dancing. The Country Seat is meant to carry on the tradition of giving young people a chance to explore the outdoors, but now they can do that through arts and nature study, hands-on crafts and cooking, and rambling around the island. Jongert and his team have managed to create a space that has the character of a big tent but also recalls the kind of mountain lodge where you might gather for an après-ski. All the while, the worn wood and assembled fragments give it a look and feel that is closer to its natural setting.

The Superuse Studio team has applied this manner of working to a wide range of buildings, from discos to private homes. They have also created parks and playgrounds. In the latter they showed that old sewage pipes and windmill propeller blades (as mentioned above) can be put to playful use. Jongert is especially fond of the windmill blades. These large vanes are shaped in elegant curves, and—because of their original purpose—they are also extremely strong and light. Their carbon fiber construction makes them difficult to recycle, so Jongert puts them to good use as places to sit or as play sculptures.

BIOPARTNER 2

Not all such forms of reuse have to look as weird or playful as those Jongert favors. Another Dutch firm, Popma Ter Steege oversaw the construction of a new laboratory called BioPartner 2 outside of Leiden, made largely out of recycled materials. When the architects began their work in 2019, they were lucky to find a "donor building" within half a mile of their building site: a 1970 complex called the Gorlaeus, which had served many of the same functions as a laboratory. Realizing that the height of the levels in one of the Gorlaeus buildings was the same as what they needed, Popma Ter Steege worked with the demolition contractors to transport to their work site as much of the steel structure as possible—as well as panels and covering that had been used for various surfaces.

Natural stone, ceramic tile, window systems, and other large-scale elements also found their way from the Gorlaeus into the new building. For smaller elements, Popma Ter Steege turned to additional sources: the toilets came from a bank that was being remodeled and the furniture from a warehouse that specialized in reused furnishings. Rubble from other building sites was placed in open metal cages to create outside walls. Felled trees from a road worksite were cut into logs and sawn into lumber for use in the new building. Bricks and paving tiles came from recycling yards.

Appropriately for a modern laboratory building, BioPartner 2 is a relatively simple block whose forms come together at right angles. From certain viewpoints, it even looks boring. The rather dull panels from the

1970 building form a tight skin hovering above a base of rubble walls. Inside, the reuse becomes clearer but looks just as clean. The carpet obviously consists of bits and pieces taken from other buildings, and a kitchen counter is formed from a stack of drawers. But even when the history of the pieces is evident, Popma Ter Steege felt it was important to blend them into the overall geometry and smooth surfaces that befit a place of controlled experimentation and research. Their achievement proves that if you work hard at designing the actual assembly of found items, you can make a building that looks new. It is no coincidence that so much of this work is taking place at a high level in the Netherlands in particular and in northern Europe more broadly.

These countries, including Germany, Belgium, Denmark, Sweden, and Norway, have the strictest and most ambitious environmental regulations in the world. They also have an openness to experimentation in architecture that is not so common in other nations. Both of these conditions are supported by their governments. The Dutch state provides subsidies at many levels for reuse, while their ever-tighter sustainable building practices standards encourage reuse. The Netherlands also has a long history of supporting architects and designers in exploration and experimentation by subsidizing study trips, research, exhibitions, and publications that test the boundaries of what we think of as good architecture.

Denmark has a similar support system for sustainable architecture. There, the Realdania By and Byg Foundation commissioned local architects Lendager in 2013 to design the Upcycle House as a way to demonstrate how materials can be reused. Though the concept single-family house is not very expressive, it shows what is possible. The structure consists of two reused shipping containers under a roof made out of recycled soda cans. The outside surfaces consist of post-consumer (reused) paper that has been pressed and treated. My favorite touch is the kitchen floor, which is made of champagne corks.

DROOG DESIGN

The methods and the aesthetics that are now common among architects there in the Netherlands, however, came together in one particular movement at the end of the last century, and it was located in the field of industrial design and furniture. In 1993, Renny Ramakers and Gijs Bakker, who were teachers and critics in that field, brought together work they had been seeing among their students and colleagues, which they felt represented a new attitude in design. They put as many pieces as they could in a truck and drove to the annual Milan Furniture Fair, the most prominent event for debuting new products. Instead of the slickest new lamp or the grandest sofa, however, they showed objects that looked like flea market finds.

A chair consisted of rags piled up, given some internal support, and cinched together with moving straps. The designer Tejo Remy's piece "Chest of Drawers" was exactly that: a loose pile of drawers that had once lived in various other chests and that were culled from a flea market and strapped together with more of those moving straps. Piet Hein Eek made a table out of scraps of wood and covered it with enough layers of transparent lacquer to make it shine like an heirloom. A chandelier consisted of a hundred lightbulbs cinched together.

The exhibition hit like lightning. Here was a completely different approach to design, one that had little use for refinement, semiotics, abstraction, ergonomics, or any of the other trends that had kept designers, wholesalers, and critics looking for the latest thing every year. The loose movement (which turned into a company in 1995) called itself Droog Design. ("Droog" means dry in Dutch, indicating the designers' matter-of-fact attitude but also their sense of humor—"dry like a good martini," as critic Paolo Antonelli remarked.)[4] They started showing up every year with new things that were old or unformed. They could be the vases that Hella Jongerius made by taping together fragments of old ones, but they could also be fragments that you had to assemble yourself: IKEA's instruction booklet and the Harvest Map combined. One exhibit, "Do" (using the English word), included a galvanized metal cube about the size of a chair and a sledgehammer. Visitors were encouraged to shape the block into either a place to sit or anything else they wanted it to be (the designers illustrated how to do so with image #2).

Droog had an immediate and far-reaching impact. One of its founding members, the designer Hella Jongerius, went on to become design director for Vitra, a Swiss furniture company, while Droog itself did projects for various other industry partners. It even influenced the late designer Virgil Abloh, who added his own touch to the idea of reuse: he could recycle an object just by printing his name on it. He started his successful fashion and design practice in 2012 by buying up an overstock of Polo shirts and printing his label's name on them.

As Droog grew, it began to pursue commissions for architecture projects, sometimes in collaboration with those trained in the field. Members also began collaborating with designers such as Rem Koolhaas, whose Rotterdam-based firm was among the most successful and influential of the first few decades of the twenty-first century. Of course, Droog was not the only source of the sort of witty and minimalist reuse that their products showed. At the same time, the designers associated with the group were beginning to emerge into their own, the Campana Brothers in Brazil were making furniture out of discarded stuffed toys and room dividers out of car antennas. Together, these designers of everyday objects were able to create a taste for recycled materials with skill and, more often than not, a certain amount of irony. The company Freitag took the attitude to mass fashion, producing messenger bags from reused wall posters and, in 2006, building their own flagship store in Zürich out of reused shipping containers they stacked into a tower.

Yet, as we've seen, repurposing everyday objects—which is what Droog, despite its own protestations, became known for—turned out to be much more difficult when constructing buildings than it was when creating furniture. It was therefore at the intersection of architecture and design—where the making of objects, spaces, and images fade off into art—that some of the most exciting work in upcycled design has taken place over the last few decades.

Dumpster diving has a long history in art and architecture. After all, the ancient use of spolia was also a form of collecting and reusing castoffs. It was practiced from the Roman era (using pieces of monuments from countries they conquered, local buildings that were no longer functioning, and even whole monuments like obelisks) through the Middle Ages (when the remains of the Roman Empire were recycled in turn), all the way to the end of the nineteenth century. This was when Gothic and Renaissance buildings were mined for pieces that could ennoble the houses of the wealthy across the Western world. You only have to think of Hearst Castle—and its fictional equivalent, Xanadu—with the press baron sitting in the middle of his perch made out of pieces of medieval castles, Renaissance palaces, and Gothic churches. Artists, meanwhile, have been dumpster diving since at least the beginning of the twentieth century.

What is more difficult to identify is imaginative reuse: those projects that not only had the spatial, constructional, and even functional qualities associated with buildings but went further to use those characteristics to explore ideas about collecting, hybrid nature, and the collapse of time and space. Ideally, such projects would have done so not through the reuse of whole existing buildings—which retain a sense of what they are and should be used for that resists easy appropriation—but through the assembly of remains, castoffs, and just plain garbage.

PERIOD ROOMS AND OUTSIDER ART

The precedent for imaginative reuse involving the setting of interior scenes is the wholesale lifting of pieces of buildings into museums and displaying them as a "period room." With roots dating back to the beginning of the nineteenth century, the vogue for preserving and exhibiting whole furnished rooms blossomed in the 1920s, especially in the United States. While at first the subjects were exclusively rooms from highly designed houses that exemplified the integrated architecture and furnishings from a particular period, soon the focus shifted to "typical" rooms that reflected how different groups of people had lived. Colonial houses were in fashion, as the period room trend became part of a larger effort to give the United States a history of its own. Evocative assemblies featured rough-hewn timbers, equally matter-of-fact chairs and tables, and works by local tinsmiths mixed with objects that early white settlers had imported from England.

The period rooms were meant to evoke history through artifacts and architecture. The rooms were reused for telling the narratives that the museums wanted to highlight. That meant that each of the objects had to work together in a recreation of the room's original use, whether as a dining room, bedroom, or place for social gatherings. It also meant that only one interpretation of one time—frozen in place by the curators—was possible. While subsequent interpretations have tried to bring into play the missing actors and stories—particularly the servants, enslaved people, and other users of these rooms who did not have the privilege to see their environment preserved—period rooms remain part of a very limited way of reusing past structures. This was dumpster diving with a selective purpose: the rooms recycled cast-off materials, but only if they told the story the divers were looking to tell.

More open to different classes and interpretations were the outdoor versions of such displays, which came to be called "open-air museums."

These were especially popular in northern Europe during the periods immediately before and after the Second World War. Sites in Germany, Denmark, Sweden, the Netherlands, and France brought together historic structures that were about to be torn down and assembled them into a facsimile of a village. Most of the structures were of a humbler nature: farm buildings, parish churches, and small warehouses. That also meant that their construction was usually not covered up by the kind of decorative patterns and finish materials that architects used in fancier structures. Instead, you could see how builders had used wood and stone largely as they'd found it and put these pieces together to create shelters. The artifacts inside were equally rough, displaying both their material and their handmade nature.

Like the period rooms, however, these outdoor museums—several of which still survive—still only told one story. The assembled buildings were meant to reflect local ways of building, and the curators of these displays either excluded those recent additions or removed them altogether. They omitted or even changed what did not seem to fit into what they felt were the national styles, and they highlighted buildings—such as stave churches in Norway—that they felt were unique to their country.

If the state and institutions assembled the remains of buildings for one purpose, people interested in telling other stories—including artists and architects—started putting together their own assemblies at the beginning of the twentieth century. Unsanctioned and often scorned by both neighbors and regulating officials, these "bricoleurs" (a French term popularized by the anthropologist Claude Levi-Strauss in his 1962 book *The Savage Mind*) put together fantastic environments in which you were unsure of what the time, place, or function might be.

You can find examples of such work all around the world; they are shrines erected along roadsides or built into walls of homes you might encounter driving through France, Italy, Spain, parts of eastern Europe, South and Central America, and South Asia. Usually made from whatever material was at hand, they commemorate significant events: where somebody died in an accident or where a deity or saint once revealed themselves to a particular person. Often tying into larger religious stories, they are also highly personal and particular, leaving you to interpret their meaning with or without knowledge of that specific narrative or occasion. While many are makeshift, others can be quite large and grand, turning into small chapels where you can enter and pray or merely shelter from the weather.

Then there are those monuments of uncertain purpose that take bricolage to a different scale altogether. One striking example from the middle of the twentieth century is Watts Towers in Los Angeles (see image #3). Constructed by the artist Simon Rodia between 1921 and 1955, the complex of seventeen spiraling structures rises next to a former railroad line, with the highest of the spires reaching almost a hundred feet. Rodia made the towers out of what he could find and what neighbors brought him. The main structure is constructed out of steel pieces he scavenged from a railroad line and twisted into spirals: they are open versions of the kind of baroque and rococo churches he might have encountered during his childhood in Italy. They also are a version of the radio and electricity transmission towers that were being constructed all around the site at the time.

The artist encrusted the metal with pieces of broken tile he collected on the street, from building sites, and, later, from people who brought him broken pottery, glass, and building materials. Rodia placed them on the structure according to a visual logic that was all his own, setting off different colors and forms against each other to make a continual pattern that leads your eye up and around the towers. Because the towers are open, views of the surrounding buildings, electricity poles, and—as you look up—airplanes approaching Los Angeles International Airport all mix in a composition that changes continually as you walk around the complex.

Rodia walked away from Watts Towers in 1955, after a heart attack and a fall made it difficult to climb its structures. He was one of several "outsider" artists who made remarkable structures out of found materials during the first half of the twentieth century.

In the 1960s, the preacher and artist Howard Finster turned his own house near the small Georgia town of Summerville into a combination church, abode, and art piece. There he built close to a thousand works of art, most made from cast-off materials; the house was also where he lived and preached to his congregation. A round tower rises above what was once a small shack, its tiers evoking a Hindu temple more than a traditional Christian church. Look carefully at the elements, and you can see that one level includes bits of porches and mismatched siding he found around his neighborhood. The religious purpose that was implicit in Watts

Tower (both through its reference to church steeples and the addition of Christian icons) and Finster's decorative impulse come together here and spill out into a sculpture site that is itself a work of art. Finster, who passed away in 2001, called it the Paradise Garden.

The modern equivalent to this sort of assembly is the Fargo House: the home, studio, and ever-changing display site created by the Buffalo architect Dennis Maher. Maher purchased the single-family home in 2009 and set about turning it into both the place he lived and a three-dimensional work of art.

The renovated (in a peculiar manner) house is a mixture of the original wood structure, which has been opened up, added onto, and carved into; domestic objects you can recognize and sometimes still use; dollhouses made up of fragments that show only bits and pieces of the domesticity they were originally meant to display; and assemblies of found objects that have no particular function and hover between being their own works of art and a continuation of the compositions all around their frames. As Maher says, these pieces do not sit still, because he actually lives in the house and uses them but also because he was always reusing his own material, adding onto and subtracting from what he had already done.

At the heart of the Fargo House is the City Wall room, where Maher built a tower made of dollhouses he found and miniatures he collected. He piled these elements on top of each other, leaving each as he found them. Their wear and tear is visible, and pieces are missing. Another assemblage in the room consists of pieces of lumber rising up through what used to be the floor of the second story, of which only the joists remain, encrusted with scraps of wood, storage elements for drawings, and what look like the remains of a model-making project. A table in front of this construction supports more fragments of models and potential building materials, each of which Maher has arranged into a composition that also implies that he could pick up the pieces at any moment and use them for another purpose.

Even the "washroom," as Maher calls the main bathroom, is not only a place to indeed wash, shower, and use the toilet but also a collection of different tile pieces laid in contrasting patterns. There are also stripped-down walls, a found bathtub, and more models, dollhouses, and construction fragments. Look in the mirror hung above the sink, and all the pieces recompose themselves in another manner. The same is true in every other room in the Fargo House: it is a riot of construction, deconstruction, assembly, de-assembly, composition, decomposition, and just piles of found objects.

A more refined—and much smaller one-room version—of such work is the installation *Before Yesterday We Could Fly*, which opened at New York's Metropolitan Museum of Art in 2022 under the curatorship of Hannah Beachler and Michelle Commander, with commissions executed by artists Njideka Akunyili Crosby, Fabiola Jean-Louis, and Jenn Nkiru. It is a combination of a period room and the kind of slightly mystical, slightly mysterious works of art described above. In fact, the curatorial team that put the display together states: "Like other period rooms throughout the Museum, this installation is a fabrication of domestic space that assembles furnishings to create an illusion of authenticity."[5] However, "unlike these other spaces, this room rejects the notion of one historical period and embraces the African and African-American diasporic belief that the past, present, and future are interconnected." The curators let themselves be guided by Afrofuturism, a movement in art, music, and more recently, film (2018's *Black Panther*) that recreates a possible decolonial Africa: one that may have existed, might exist in the future, or may still be there, hidden away somewhere either on the African continent or in the diaspora of Black people throughout the West.

The Met's installation refers to Seneca Village, a nineteenth-century settlement of Black people in Manhattan that was eradicated in 1857 to make way for Central Park. A clapboard house that looks something like a dwelling of the village is there only in part. It could be either in ruins or under construction—a relic of the past or the beginning of another future. Within the structure, the curators display a wide variety of artifacts scavenged from the Met's holdings of African art, African American art, and other collections. The museum pieces are interspersed with commissioned work that begins to trace what Seneca Village might have been, what it might be some day, or what it still might be if you look at Central Park with new eyes.

Works such as *Before Yesterday We Can Fly* are period rooms with a purpose. Other recent examples recreate supermarkets to comment on our addiction to consumption but also to tie into a certain nostalgia for the brands of our childhood. There are also temples to fictional religions. These spaces combine the period room's ability to give new meaning to collected artifacts with the ability of artists and bricoleurs to tell not just one story of one time and place but a continually changing reality. They transform dumpster diving (even if the trash bin is a museum's collection) into world making.

Then there are those artists who create worlds, or at least temporary environments, completely out of trash and leftovers. In 2012, the French artist Pierre Huyghe produced an example of such a way of renovating by doing almost nothing.

Huyghe's untitled piece, whose official description was "Alive Entities and Inanimate Things, Made and Not Made/Dimensions Variable," was sited in a section of a public park in Kassel, Germany, where maintenance equipment and refuse were stored. Huyghe accepted much of what he found there, making only some rearrangements. He covered piles of leftover plant material with enough earth and sod that you could sit on them. The flower seeds that then were added to the mounds came from Afghanistan, which was a focal point of the larger art fair, Documenta, which is held in Kassel every ten years and which this installation was part of. Concrete pavers, usually used to make walkways, became seating elements around a pile of more pavers and dirt, offering a place for informal lectures. A found sculpture of a recumbent woman had her head covered with an active beehive. The only new piece within the work was actually a living creature: a stray dog the artist encountered, befriended, and somehow convinced to let him paint one of his legs a bright purple.

Walking into *Untitled* along with a crowd of other art seekers, you had no sense of where the art began—the space was uncertain in its boundaries and its time. You certainly were aware that you had entered a reshaped version of the park by the time you were sitting on the pavers listening to a poet declaim while the purple-legged dog sniffed at you and the bees buzzed around the sculpted woman, but as the crowd dispersed and the dog wandered off, the space faded back into its surroundings. It certainly did so at the end of the summer of 2012, when this edition of Documenta ended.

THE ARCHITECTURAL POSSIBILITIES OF DUMPSTER DIVING

This confusion about where the art or architecture starts and stops is inherent in work that is assembled out of elements that have supposedly outlived their usefulness. If they were discarded, why are we looking at them or using them again? If they didn't work anymore or were not beautiful, why should we care for them? We know that dumpster diving is a necessity for some and a pursuit for others, but how is it architecture or art?

The first answer to these questions is that we cannot afford to make anything new when it involves extracting from and fouling our planet. We must therefore preserve and reuse, and that includes what we might think of as junk. The second answer is that we must also remember and tell stories that help us create continuities in time and place, even as the world all around us keeps changing. Discarded elements, out of context and reframed, are in a way the poetry of such stories. Isolated and newly assembled, they speak with great clarity.

The third answer is the need, from a sustainability standpoint, to salvage and reuse as many parts of buildings as possible, even if they have been thrown out. The strategy of dumpster diving that comes out of that reality creates a new kind of beauty, one that depends exactly on its critical distance from the aesthetic system within which the material was originally created. If a piece of steel is no longer a hidden support for a building face, then it is free to support and stand for something else. It usually cannot do so alone. It is the combination of various pieces from different places and times that lets this new work of art or architecture appear.

Architects are still trying to figure out how to be good dumpster divers. Even the best buildings I have cited above still have a tame quality to them, as if the architects were still trying to make them conform to the standard design rules. If architects look at the work of any artist who can explore the possibilities of cast-off materials without such constraints, they may be able to create architecture that is not only good for the planet but can open other possible worlds for us.

PART 2

TRADITIONS

CHAPTER 3

EPHEMERAL ARCHITECTURE

THE SHRINE OF ISE JINGU

Every twenty years, you will find carpenters at work in the woods in the middle of Japan, about a hundred miles east of the old imperial capital of Nara. Using tools that were first developed almost two thousand years ago, they are replicating either one of two shrines that stand about four miles from each other or the bridge between them (see image #4). The design of these structures is as old as the tools the workers use. There are no nails, and they use no material other than wood and thatch. This complex, called Ise Jingu, or Ise Grand Shrine, is dedicated to the sun goddess Amaterasu of the Shinto tradition and is to this day still one of the holiest places of worship in the country.

According to the shrine's own mythology, the daughter of the ruling emperor established the worship at the site in 4 BCE, even though the architecture reflects the style of what is known as the Kofun period, which comes somewhat later in history (300 to 538 CE). Ise has become known in the Japanese imagination as the mythical place where the sun goddess alighted and then decided she would make her home there among the forests and streams in this isolated part of the country. Built of local untreated wood (at least originally; now the material comes from all over the country), the shrine structures consist of pillars that rise up around a raised platform to support a thatched roof. Around each building there is an enclosure large enough to allow for the construction of new shrine while the wood of the existing building slowly deteriorates over the course of two decades.

The shrines at Ise are both ancient and new: they are no more than twenty years old by definition, but their exact form, material, method of construction, use, and placement have remained the same for centuries. To what era do they then belong? Not to any in particular but to a cycle of time all their own, in which their wooden parts establish an order and meaning before they slowly fall apart over the period of approximately one human generation. The architecture binds each successive community to those of the past and to the place.

Ise is an isolated site, and the importance of the shrine complex is immense: the initiation of each new building is attended by the Japanese emperor, who gains a part of his legitimacy from his visit. Only the emperor can enter the holiest part of the inner shrine, where Amaterasu's mirror is kept; the general public cannot proceed beyond the main gates. The buildings have no relation to contemporary culture or life, let alone economics. They exist only because this particular society is willing to continually invest in their reconstruction so that it can found its political and cultural structure in a place and architecture shrouded in myth, yet very real.

The Japanese approach to the traditional forms that Ise represents is an alternative both to the worship of the old and to a modern throw-away culture. Instead, this architecture shows that when a culture believes in something, they do everything they can so that it is maintained, and that means rebuilding it continually. Without such an agreement, the buildings would fall into disrepair.

In the cultures of Europe and the Americas, such a structure might be reconstructed or fixed up as a reminder of what it once was, creating a three-dimensional representation of a shrine as it once functioned. The resulting building would serve as a reminder of the original, rather than functioning in the manner for which it was constructed. Alternatively, such a structure could just disappear, leaving only its memory in stories—whether written and oral, painted or photographed—to keep some aspect of it alive.

What Ise Shrine has come to represent, then, is a challenge to the notion that we should design buildings to last, as enduring monuments that fix our values, beliefs, and ways of doing things into form. Instead, the Ise model is one in which a building lasts only as long as a society's commitment to it and everything it stands for. The building, in turn, is constructed to last only for that use and period of commitment. A generation is a pretty good measure of time to gauge a society's devotion to an architectural form because it allows the next generation to understand its value. This approach requires that we only invest as much time, effort, and materials into the buildings as is appropriate and that we design the structure so that it either goes from earth to earth, dust to dust, or forest to forest, or is repurposed, reassembled, or otherwise reused.

Ise Shrine is also a challenge to the debate between those who believe that the most sustainable and appropriate way to build is to make a structure last and those who say that we should commit only minimal resources to a structure. While the former attitude has long been central to the thinking of most architects, the latter is now gaining traction. After all, our society and its citizens change much too fast for a building to function for long in the way its design was originally intended.

What is it that architecture can and should provide in that sense? I think it is two things: a modularity and flexibility, and a way of creating meaning that binds us together and frames us while using the least possible material and forms. To do so, architecture needs to be not permanent but ephemeral.

Ise Shrine, it turns out, is only a highly refined and condensed approach to architecture that offers a possible alternative to the wasteful and monument-focused ways of the Western tradition. It shows how a minimal investment of material, if repeated and refined, can create a huge amount of meaning. It also shows that breaking down a built structure into simple and standardized components—the columns, beams, and connecting elements—does not necessarily mean that the result becomes bland through neutrality. For such a structure to work, however, it must be made more specific but also scalable and adaptable. Beyond those requirements, it also has to be able to carry meaning by being part of a larger complex of cultural construct. The building has to be expressive, carrying memories and values with it, allowing us to either commit to those traditions or break them open.

In the case of the Ise Shrine, the baggage is rather heavy. Despite the myth that surrounds it, in historical fact its construction was a way for the Yamamoto, one of the many third-century warring factions vying for control of the country, to anchor their claim that they were descended from Amaterasu and were thus the nation's sole and legitimate rulers. Once they had succeeded in turning themselves into the imperial family, they then built the shrine as a pilgrimage point, but also as a way to connect origins to destiny: based on vernacular and perhaps even construction techniques from their native Pacific islands, the shrine transformed from its humble origins into a highly refined form framing a choreography of power.

The construction technique that Ise embodies is surprisingly simple and universal. The buildings rise on their poles out of the ground with little foundation or transition. They support a deck held well above the ground, on which the actual enclosure is built. The thatched roof is a simple gable whose beams crisscross, with their ends extending beyond the edge of the enclosure. The front beams are covered in copper to symbolize the sun. The inside is, by all accounts—as only the high priest and the emperor and his family ever enter—equally simple.

Ise Shrine is thus what architects have long called a "primitive hut": the simplest abstraction of wood branches and leaves interwoven to create a shelter. Western theoreticians as early as the eighteenth century speculated that our tradition of more permanent and monumental structures, which they postulated as originating from classical Greek temples of the fourth and fifth century BCE, were translations of that forest language into stone. Though the argument has a certain amount of intuitive structural logic, as those temples look as if they could have originally been built of wood, there is no empirical evidence of that development.

Ise, instead, remains as wood. It gestures to both simple functions and to a mythical point of origin: the structure resembles not places of human habitation but barns that are lifted off the ground to prevent pests from raiding stored food while also keeping the structures safe from floods. It is thus an abstraction not of spaces sheltering human life but of wealth represented by surplus grain.

The specific form that the shrine took is one that probably came to Japan with the Polynesian invaders who established the country's ruling class. The resemblance of the shrine to the structures still found on many Pacific islands is, in fact, remarkable. The shrines are as much a built embodiment of imported and imposed power as they are of native traditions, but the way they also resemble the humble structures of everyday life makes that alien nature fade away.

The modular system into which the Ise Shrine was refined around the seventh century has a different source. It probably came from the larger cultural and political force to the west, namely China. It closely resembles the Chinese modular system of *dougong*, codified in the sixth century, in which a single bracket or connection between a vertical post and a horizontal beam is elaborated into temples, palaces, or other significant structures. That system, however, is resolutely reserved for important buildings and was not meant to accommodate places of daily life.

Similarly, Ise Shrine is not a good reference point for making houses or office buildings. It is a model, however, for modularity, simple materials, and open spaces.

EPHEMERAL BUILDINGS IN MODERN JAPAN

The Ise prototype is, of course, difficult to carry out in urban settings, but what has remained from it is the idea that, as much as possible, interior spaces should be simple, reconfigurable, and modular. Many modern Japanese dwellings, even if they are completely Westernized, have at least one "tatami mat room" (a room defined by straw mats arranged to cover the floor completely) where the dwellers can live or hold rituals in the same type of space in which their ancestors would have. The spirit of Ise Shrine also speaks through larger structures by contemporary Japanese architects—in particular those working in a lineage started by Toyo Ito in the 1980s, developed further by his one-time collaborator Kazuyo Sejima, and now carried on by a group of Sejima's former apprentices and students.

The Japanese tradition argues for architecture that is reduced to its simplest elements. These pieces are rooted in the local conditions, materials, and history. They are modular and can create reconfigurable rooms. These spaces, defined by screens, can be opened and closed easily and are minimal in their furnishings, while being laden with meaning because of the traditions out of which they come.

Such an approach to making spaces is, especially these days, limited. Sejima and others have tried to create not only houses but also dormitories and cultural spaces that turn the modules of tatami mats and the simplicity of traditional forms into three-dimensional puzzles, but these examples are few. Moreover, they are hard to find in the sea of concrete and steel that makes up most cities there, as elsewhere. What we can learn from studying that approach, however, is valuable. Not only can simple be beautiful; it can also have meaning and be flexible, as long as it is carried out in a manner that evokes history and provides enough concrete points of attachment or framing (affordances) to let us lead our lives within them.

Reuse, in other words, can be a question not just of inhabiting and adapting existing structures but of committing ourselves to using, maintaining, reconfiguring, and even rebuilding the simplest possible frameworks. That means not just finding functional ways to use what is there but also investing ourselves in the meanings that come with the forms and letting those forms die and be reborn in new ways. This is an important answer to the monumental tradition so central to Western architecture. We need anchors, reference points, and places that connect us to the past, but they should be as minimal as possible, able to be used and extended in other ways, and last only as long as our commitment to them.

THE EPHEMERAL BUILDINGS OF AFRICA

Japan is by no means the only place where this anti-monumental and modular approach is evident. In Africa, for instance, some of the grandest structures are made of mud and have to be rebuilt periodically. Most famously, the libraries, mosques, and palaces of several former capitals of the Islamic empire—as in Ghana and Mali—have been continually reconstructed since at least the thirteenth century out of mud that is compacted, strengthened with wood beams, and covered with a stucco-like painted covering to preserve the vulnerable mass.

After the conquest of the area by a group of Islamic tribes in the thirteenth century, structures were built to replace and refine mud construction techniques. Timbuktu's Grand Mosque, founded in 1327, has been rebuilt and expanded several times, each new layer burying the earthen structure below its contours. The libraries have remained more static, although they have yet to be studied by the kind of archaeology that has pinpointed the exact history of the mosque. What is most important is the fact that the rebuilding continues to this day—not just after bouts of violence destroy structures, as happened when Islamic extremists took over Timbuktu in 2012, but also as a normal activity of that society. Even with a harder covering, the earthen structures deteriorate over time, and the community must continually rebuild them to maintain their integrity.

The result is a collection of forms that are both ancient and grand in their appearance. They recall an unbroken tradition of learning and faith, and they are an integral part of their landscape. They rise out of the edge of the desert, demarcating the crossroads that made Timbuktu so wealthy, both monetarily and culturally. They also offer a human-made answer to this transitional geology and climate, standing as steadfast monuments between continually shifting desert sands and more tropical landscapes, and thus forming havens and markers for caravan journeys that start and end there. These monuments are of their place but out of time.

You can find similar earthen structures throughout North Africa, the Arabian Peninsula, and into Iran and Afghanistan. They continue to shape and focus communities, but they only endure where communities have committed to them. For the few remaining intact collections of adobe or mud villages and monumental structures, there are countless more that have disintegrated back into the land. In the United States, we can still find remains of Hopi communities, like the one in Taos, New Mexico, which has been rebuilt out of adobe for centuries (unlike many other Hopi and Anasazi communities, which have disappeared) and, we hope, will continue to persist.

THE INSTANT CITY OF KUMBH MELA

An even more ephemeral construction is the vast city that arises in central India every ten to twelve years for the Hindu festival of Kumbh Mela, which celebrates both community and individual sacrifice. The festival takes place on a shifting floodplain at the confluence of the Ganges and Yamuna Rivers. It is rooted in ancient traditions, though its current form dates only to the middle of the nineteenth century. In 2013, the last well-documented occasion of the Kumbh Mela, over seventy million people lived in a city that only existed for the month of the festival. This was no mere tent encampment. Focused on a central, sacred area marked by a flag, the Kumbh Mela sprawled around the banks into a complete urban conglomeration that included transportation hubs, health facilities, and administrative areas made out of tents and sometimes reused wood and corrugated metal. An infrastructure of floating bridges, raised steel walkways, and sewage treatment facilities tied the city together and largely consisted of reused components from other sites. All of this lasted only as long as the millions were present (you can see what it looks like when it is in use in image #5).

This instant city must be built fast, as the exact contours of the river banks remain unknown until right before the festival. Within a few weeks, all the construction happens on the site, and the pilgrims arrive to bathe themselves in the Ganges, attend lectures by various gurus, and engage in social gatherings. The city has a grid of streets, though it is fractured, with the various squares and rectangles that organize the tents and other structures coalescing around certain temporary temples or responding to contours of the site.

The architect and critic Rahul Mehrotra argues for the Kumbh Mela as an alternative model for urbanism. In his discussion of the phenomenon, he remarks on three characteristics of this instant city that are worth considering. The first is reversibility: "The Kumbh Mela," he says, "challenges the idea of sustainability, by engaging us to think about urban design as a reversible operation."[1] Instead of trying to anticipate how people will live, work, and play in the future, then investing in immovable infrastructure, Mehrotra argues, we can think of our cities as places that are adaptable and ephemeral, maintaining their form only as long as we need it. The materials that are used to make the festival happen, moreover—tents, scrap metal, wood, and infrastructural components such as bridge pontoons and toilet cabins are almost all reused and recycled, leaving few marks on the landscape.

The Kumbh Mela's second notable characteristic, according to Mehrotra, is its openness—not only in its form but also in its combination of minimal top-down planning and significant local improvisation. This, Mehrotra says, should help us to think of cities as "open systems" in both time and place, rather than the static and closed forms of most of our urban environments.[2] In such a system, the architecture that shelters and frames us appears, changes, and disappears in a quick succession of scenes, with the stage sets for each appearance repurposed for the next.

That leads Mehrotra to point out the third thing we can learn from the Kumbh Mela, which is its "nuanced temporality"[3]: it is a city that appears, disappears, and reappears, preserving traditions and forms while also changing every time and even during each iteration. This is an approach that we certainly can use in responding to disasters or other dire events, but it might also be a way we could think of our urban environments in general—as acts of continual reuse and renewal.

Whether it is Ise Shrine, the monuments of Timbuktu, or the Kumbh Mela, it is worth noting that these structures are no less grand in their scale and appearance than many of Europe's cathedrals, palaces, or other monuments. Moreover, evidence is increasingly coming to light that there were many more such ephemeral structures around the world and at many different times, but, as colonization wiped out local economies and the power structures that controlled them, the commitment to continually reusing and rebuilding these structures disappeared starting in the sixteenth century.

In fact, even within European history there is a strain of ephemeral monumentality. In the Middle Ages, movable royal and aristocratic households or courts would travel between bare-bones castles or even encampments, deploying a domestic and political armature in the form of furnishings and cultural signifiers—such as banners and art works—wherever they alighted. That tradition, in turn, transformed into the highly ritualized "masques" of the sixteenth through eighteenth centuries, in which important occasions—such as royal weddings, the arrival of ambassadors, and coronations—were restaged as mythological events with both rulers and subjects participating. There were also other celebratory fairs without the direct involvement of the aristocracy, such as those that appeared at the end of Lent or right before Christmas. They were free and temporary places that stood outside of the strictures of the city and its laws.

Throughout recorded European history there were also encampments of armies, which brought large groups of people to one place and organized them in tents and other temporary structures in a rigid, purely functional layout. These traditions perfected the basic building blocks of what could be a different mode of inhabitation, one that relies on moveable, reusable structures and a light occupation of the land.

BURNING MAN, COACHELLA, AND MODERN NOMADISM

It was not until the mid-twentieth century that many of these models merged into something that combined and heightened their civic celebratory aspect: gatherings around rock and roll and, later, other forms of music. Examples include Monterey Pop in 1968, Woodstock in 1969, and annual festivals in Europe such as Glastonbury and Pink Pop. These gatherings were meant to be festivals of art and music and drew hundreds of thousands of people. They had a focal point (the main stage), often subsidiary stages, and areas mapped out for craft fairs, food, health care, and sanitation, as well as open areas for camping. Music might be the draw—the religion of our day—but the act of pilgrimage and gathering for self-definition is just as important.

The tradition continues to this day all over the world. Not only the music has changed along the way—so has the nature of place making. The most famous "instant city," Burning Man—which appears in the Nevada desert once a year around the ritual of burning a large wooden effigy—is designed as a perfect circle around a central core, which is how many such ideal cities have been envisioned by architects since the Middle Ages. At Burning Man, concentric circles of encampments—consisting of tents, RVs, and other temporary structures—focus on the Man, or effigy, and sectors are laid out with shared services. Within each sector, visitors vie to create the most elaborate and expressive forms of artwork, including sculptures, performances, and structures to inhabit. The organizers, working from a nearby base camp, make sure that nothing is left behind once the event is over and that the desert is restored each year. The problems of such an approach became clear in 2023, when heavy rains flooded out the whole event. Living light on the land also means being subject to the forces of nature.

The gathering that fascinates me most is the music festival that occurs for two weekends every year in Coachella, California. It is located in the middle of seemingly endless red-tile-roof residential subdivisions on the site of a former racetrack. As such, it acts as a palimpsest of sprawl, connected not only to roads, electricity, water, and sewage grids but also to the openness and undefined character of space there. The central area is less a perfect piece of geometry than a kind of central park, dotted with sculptures that the organizers commission each year, and ringed with stages that act as venues for different kinds and scales of performance. It is thus more varied and less centrally focused than most other such events.

Coachella has a food market as well as restaurants, toilet and shower facilities, and stores. It also has meditation spaces, exclusive clubs, and a healthcare clinic. Behind the proverbial screens, security operates as invisibly as it can out of trailers. At the periphery of the whole complex you can find a tent city where habitations range from the most basic to luxury glamping. Parking lots and logistical hubs that ferry buses and taxis in and out of the zone complete the temporary city. The organizers also commission architects to create temporary civic structures that serve as landmarks in the sprawling terrain, provide shade and shelter where people can gather, and, of course, become selfie backdrops.

What is most remarkable about these festivals is the way they create an instant community. People arrive from around the world, set up their tents, find their way to the central area, and mingle in a secular celebration of something they all value. The similar look not only of the people and their clothes but also of their temporary dwellings and the civic monuments around which they gather is defined by class and privilege to a certain extent (tickets to these events can cost hundreds of dollars) but also by the ever-changing culture of those there.

A tent designed for one of the Coachella stages by DoLab (a combination design, construction, and event company) demonstrates some of the possibilities of the festival. DoLab was founded by three brothers whose parents toted them around to Grateful Dead concerts and hiking campgrounds when they were growing up. The boys then attended architecture school in the late 1990s, just as a renewed interest in the possibilities of tensile structures was being developed, further fueled by new computer visualization and modeling tools. Out of this background, they came to focus on the design of temporary, demountable tent structures. For Coachella every year, DoLab designs a new version of their most visible signature structure: an arching canvas semidome, big enough to house up to a thousand people and erected out of crisscrossing metal pieces. The color of the dome is different each year, and DoLab keeps refining and adjusting the form and layout. Grand and sweeping in its reach, the dome is also highly legible: the way it is constructed and erected is immediately understandable to anyone who sees it, allowing attendees to feel a connection to it.

Coachella makes me wonder if there might be a different way to think about how we inhabit the earth as communities. Traditionally, we have gravitated ever more toward urban centers and their fixed facilities. Services have to be brought into these places, waste taken out, and the buildings themselves have to be continually adapted according to changing needs, technology, programs, and tastes. Some cities and towns fall out of favor, leaving gaps of unused space or structures. Others grow so large as to lose the sense of their original focal point—be it a fortress, church, bridge, the end of a mountain pass, or a factory—and become places only by association.

Now, however, nomadism is on the rise again. For millions of people, that movement is not a choice. Human-made and natural disasters, which often converge, are displacing more and more people, and the prospect of more severe climate events, as well as the ever-growing disparity between those with resources and those without, will make such displacement a reality for an increasing percentage of the world's population. Refugee cities are thus also growing, and they are, on the whole, dreadful places. Some of them become semi- or even fully permanent, like the ones where many Palestinians live in Jordan and Syria, or the huge South Asian and Latin American slums to which people from the countryside flock in search of economic opportunity.

On the other end of the spectrum, the "operators of symbolic logic"—as the lawyer and former Commerce secretary Robert Reich calls people who produce not goods but flows of information, logistic, or ideas—are no longer tethered to one place.[4] In a movement enhanced by the global health crisis of 2020–2022, many are now working at home. That home can be anywhere if there is a good Internet connection and maybe a decent transportation hub nearby. Thus, instant communities of digital expats are arising in what used to be farm communities or resort towns. Bozeman, Montana, with its proximity to spectacular recreational opportunities, has become what locals call a "Zoom Boom Town."

At the fringe of that movement are those who choose not to live in one place but rather stay in touch with a community that is important to them—which may be based on work, music, sex, religion, or other strong social binders—by traveling around the world to meet and collaborate with this continually changing cast of friends and associates. The members of this "vanlife" or WeWork-focused movement (named after the now-bankrupt company that was one of the world's largest lessors of office space in the late teens of this century) come to a temporary halt periodically where there are enough of those people with shared interest in one place. Tony Hsieh, the founder of Zappos, had until his untimely death in 2020 created the most successful—albeit temporary, given the nature of the movement—of these encampments near his headquarters in downtown Las Vegas, reusing parking lots and vacant building lots for an encampment focused on a screen where he showed movies.

What if we can realize the need for community in a way that is not permanent or fixed but temporary and moveable? What if we can use the technology we already possess to make mass-produced temporary structures—from tents to trailers and campers—into better places to live? What if we can learn from the likes of the Kumbh Mela how to create infrastructure that is light on the land and can be reused, while also providing an arrangement around civic, cultural, and religious gathering points? And what if those structures can be rebuilt in a manner that is recognizable and revered, like the shrines at Ise, but, like that pilgrimage site, in different locations?

The possibility of a temporary, mobile urbanism is out there. It can be designed to make life better for the millions who are looking for a place to belong, whether they are forced to do so or want to find such a community. It can be light on the land and continually recycled. It can also be a place that celebrates the vital and continual work of making a community.

CHAPTER 4

GHOST ARCHITECTURE

Building from Accumulations of the Past

THE EVOLUTION OF 500 CAPP STREET

It's one thing to renovate a building. It is another to bring the past back to life. At 500 Capp Street in San Francisco's Mission District, the walls glow. They have been varnished and partially stripped of layers of wallpaper, while the bare areas have been painted. The wallpaper that remains has been covered with beeswax, leaving it to glow in the soft light coming in through the front hall. The strangeness of the house does not end with its surfaces. A bundle of brooms leans against the wall, their days of sweeping long gone; they now merely remind us of that work and invite us to enjoy their forms and textures. A gash in the wall commemorates, according to a plaque affixed next to it, that this was where a safe, on its way to its disposal, bashed a hole in the wall—for the second time. Everywhere you look, what you might think of as junk forms a composition, and what looks like an unfinished or abandoned surface gleams or presents a picture of long-gone habitation.

The house at 500 Capp Street was modest when the artist David Ireland purchased it in 1974, intending to make it his home and studio. When he began to renovate the home, his method of preservation and re-inhabitation yielded a new approach. It is one that makes history part of a structure's continual present. Around the world we can now find architecture that is an act of making both personal and social memory present by preserving, highlighting, and recomposing fragments of construction and furnishings. The real aim of this kind of renovation is not to clean up the building and make it easier to use, even if it does achieve those ends, but to let whatever we do in the structure become part of the accumulation of textures, objects, and images that we find there. It is a celebration of the continuity of life, the preservation of its joys and miseries, and the creation of a cocoon from which we can venture forth into the future.

That is not to say that Ireland started this approach to imaginative reuse or was responsible for it spreading around the globe. But 500 Capp Street, which today is a publicly accessible art installation, is one of the earliest and still one of the most radical examples of such an approach.

The work he performed on 500 Capp Street did start out as a more or less normal renovation. Ireland was trained as a carpenter, although he had also worked as an insurance agent and as a safari guide before obtaining a Master of Fine Arts degree in studio art at the now shuttered but once hugely influential San Francisco Art Institute. He came into his chosen field at a time when artists were questioning what they did more than ever, creating what one critic called an "expanded field" that ranged all the way from sculpture to architecture and landscape shaping to performance and political activism.

Ireland was working at a very particular time: both in art and architecture, the long-held idea that there was an ideal that the maker should pursue—and that this ideal should be either aesthetic, social, or both—had come crashing down in the 1960s. The mid-century's grand architecture of glass skyscrapers and public-housing towers—along with the industrial and war complexes that had made them possible—had turned out to be violent, inequitable, unsustainable, and often just not very practical or pleasant. The pursuit of the perfect work of art that many artists had pursued—especially minimalists and lovers of abstraction from Barnett Newman to Donald Judd—led to an ideal of almost nothing, to systems that folded in on themselves, and to things that called themselves art but were incomprehensible.

San Francisco's Mission neighborhood was a mess. It was cheap because it was near a ghetto, and the city itself was dirty and, some felt, dying. The worlds of art and architecture didn't seem to have many ideas on how to picture or frame this environment of economic malaise, pollution, riots, and social injustice, let alone how to make sense of it or even make it better. As it turned out, the first step was to look at what was around, where it had come from, and make sense of it as something found. Out of archaeology, collecting, and framing a collection of new orders, something could emerge in which an inhabitant could find themself at home in a physical place with a complex history evidenced by the structures there.

The project at 500 Capp Street stands for the moment and place at which many artists and architects realized the dreams of making a better world were dead, and that we should instead inhabit and enjoy the ruins. Beauty could be found in what already existed, meaning in the past, and possibilities for a better life exist in scraping, marking, opening, and assembling the leftovers.

Ireland placed himself in that vast field of possibilities by concentrating on how he could tease meaning out of the preservation or manipulation of found forms. Before 500 Capp Street, he first made a name for himself in the late 1960s and early 1970s with various works. He poured ninety-four pounds of cement on the floor of a gallery and then cleaned the mess up, leaving only the traces of what is usually the beginning of construction. He also collected everything from African sculptures and animal skeletons to leftovers from construction sites and bits of concrete he shaped into balls and wads for later use. What interested him about the objects he collected and sometimes combined into art works was their material, feel, texture, smell, and the stories they evoked. He called himself a "reader," and one museum director, Jock Reynolds, thought he was "somewhat of an archaeologist or folklorist."[1]

Ireland started his work at 500 Capp Street simply enough, by fixing the sidewalk in front of the house and then clearing out some of what he found within the walls. Before long, however, Ireland's attention and focus seem to have wandered from cleaning and clearing up to preserving, highlighting, and curating the material he had encountered when he bought the house. He started reading not only the scraps of archives about the house, but also reading the remains themselves for clues about the lives that had been led there. He started telling stories he made up out of what he encountered. Those narratives were not literal but composed of the walls he scraped, fragments of building materials, parts of furniture and even brooms that were there, and incidents he recorded, like the safe crashing into the wall "for the second time."

The artist regularized all this activity into a series of art works that he called *A Portion of: From the Year of Doing the Same Work Each Day*

(1975). He used the word "activations" to refer to his acts of revealing and preserving—a term that other artists of the time were using to describe work that was less about making pretty pictures than social critique.[2] Ireland pried open floors, let windows hang without surrounds in walls stripped of covering, removed carpets and then displayed pieces of them, and kept finding new ways to mine the Capp Street house for what he called "scientific specimens." Soon, those specimens included jars of sawdust and sheets of wallpaper. They were proofs, he felt, of the personalities of particular former inhabitants, who were now present as ghosts in the house's machinery of construction and decoration. He made some of the found artifacts into sculptures—works that stood by themselves, that you could buy, and that followed at least some of the rules of traditional composition and beauty—but to really see Ireland's art, you had to visit the "old painting," as he called it, that he inhabited (you can get a sense of what he made in image #6).

Visitors came, at first from the local art scene and schools, but soon from around the world to see this curious construction of art and architecture. The house was not without precedent in the art world. In the 1920s, fifty years before Ireland began work at Capp Street, the artist Kurt Schwitters had turned his own home in Hanover, Germany, into a livable sculpture he called "Merzbau." It, too, consisted of found fragments, although in Schwitters's case they were bits of trash he'd picked up off the street or pages he'd torn out of newspapers.

If Ireland was working in the tradition of collage and assemblage, he was also evoking another one—that of collecting. As many critics have pointed out, creating an assembly of objects that surrounds you with beauty, strangeness, illusions, and allusions is the most extreme example of how you can make a world for yourself in which owning and controlling things is at the center of how you are judged or judge yourself. It shows that not only can you afford those objects but also that you have cultivated the knowledge and taste to pick the right ones and put them together in the "right" way—whatever that term might mean. The collector is somebody who amasses art or other items but also surrounds themselves with the results of that passion to such a degree that they become part of the resulting assembly. Their life, defined by accumulated possessions, is transformed into a form of art.

THE ART OF COLLECTING

The collector is, on the whole and with the exception of those who collect very high-value objects, a middle-class phenomenon. They acquire things, as a middle-class person does, and those objects are usually for use or decoration, helping situate the status of somebody who neither comes from a long line of aristocrats nor defines themselves by land they work or the objects they make. The collector, however, takes their collections beyond what most people think is a functional activity. They build up assemblies of things that might once have had use, like porcelain or silverware, or toy trucks or manuscripts, or paintings or furniture. In the collection, the worth of these objects becomes not their utility but their specific characteristics. The pieces become art and part of an art collection. Even objects originally made to be art lose their special status when a collector puts them together in a series of works that overwhelm each specific example. The collection then becomes the point: how many, of what quality, and of what rarity its components are, and how complete the whole series is. Collection almost always tips to, or has aspects of, mania. The more you have, the more you have to have, and the items surround you, penning you in with their presence. At the same time, the collector also becomes the collection, which is a portrait of them, communicating their values, beliefs, and, quite simply, worth and knowledge, while also reflecting themselves back to confront their own body.

Starting in 1794, more than a century before Schwitters was at work in Hanover, the architect Sir John Soane made his London house at 33 Lincoln's Inn Fields—a neoclassical structure he had designed himself—into a museum of architectural fragments. The fragments were packed so densely around the living areas that they merged with the architecture. He collected fragments from ancient temples, bits of more recent architecture, and images of buildings. His collection was part archaeological and part three-dimensional source book to which he could refer in designing his own buildings. Over time, the difference between what he used as inspiration and what he designed became smaller. Soon the house and the collection also merged, with fragments becoming part of the construction of the rooms. Display mechanisms—like racks to show drawings—overwhelmed the walls from which they hung and the floors on which they stood, making those surfaces into deep and moveable edges. The rooms themselves became more ephemeral, as if they were stage sets or scrims. His breakfast nook was a spot of space over which a "napkin dome" ceiling hung as if by magic.

Sir John Soane's house became a hybrid of a museum, a home, a three-dimensional catalog, and the picture of an obsession with one form of art—namely architecture—that so overwhelmed the inhabitant that he completely disappeared into the assemblage: we remember him today above all else for the work at Lincoln's Inn Fields, and even during his lifetime visitors described him as almost buried in his accumulations. Like Ireland, Soane declared his house a pilgrimage site that could be visited in perpetuity to continue the history and interpretation of its artifacts.

MATERIAL CULTURE

The idea that we can understand past ways of life by looking at the artifacts people used and the places they inhabited is a mainstay of an academic pursuit called "material culture," which seeks to understand history not through great deeds done but through lives led and objects left behind. The field first arose in Germany in the middle of the nineteenth century, but it was the United States where, especially in the guise of a field called American Studies, it became most widespread, starting in the 1960s. In a country that had to invent and define itself, the way of doing that—with industrial artifacts, house styles, grand styles of architecture, and other means—provided better examples of the nation's character than portraits of great men and women or battles fought. American collectors—including the archaeologist Henry Chapman Mercer (founder of the Mercer Museum in Doylestown, Pennsylvania) and patent medicine king Albert C. Barnes, who turned his suburban Pennsylvania home into the Barnes Foundation, later relocated to a new building in Philadelphia—had, in the century before the rise of American Studies, placed farm implements next to paintings and sculptures and displayed them as part of a pattern of portraiture of this country.

The American focus on materialism also led to the construction of period rooms, which, as we saw in chapter 2, were a mainstay of both American and European museums from the end of the nineteenth century until the 1960s. You could go to the Brooklyn Museum, starting in the 1920s, for instance, and see reproductions of colonial dining rooms, complete with the painted wood and patterned wallpaper favored by well-to-do households of the era. Ironically, those period rooms had fallen out of favor by the time Ireland started working on 500 Capp Street, with most of them relegated to storerooms or even discarded.

Concerns about authenticity and diversity of representation in such exhibits fed into divergent attempts to picture and reconstruct history. On the one hand, there was the Disneyfication of display, starting in the 1950s, in which that company and its imitators could create seamless and completely artificial evocations of other places. Disneyland's Sleeping Beauty Castle, more commonly known as the Magic Castle, and Main Street U.S.A., both of which opened in 1955, evoke different historical periods and have been revised many times since, with little concern for actual history, use, or people. On the other hand, there were artists who cast a critical eye on such efforts, creating assemblies that were purposefully fake or questioning. In the 1960s, pop artist Claes Oldenburg made storefronts in which he displayed crudely handcrafted versions of everyday implements, as well as of some of Disney's most hallowed characters. In Germany, starting in 1970, Joseph Beuys fictionalized his own experience in the Second World War into artworks that consisted of felt suits, sleds, and blocks of rendered and congealed animal fat, either presented by themselves or assembled in variations. These were meant to portray the sled on which he had been carried out of a crashed airplane, the felt in which he was wrapped, and even the plane itself. He extended his stories into performances and into collectable artworks that consisted of pieces of these artifacts and other leftovers, which he imbued with meaning and either framed or exhibited in jars.

Most radical were those artists who flipped the act of collecting by not assembling their artifacts but instead by seeing their art as a way of discovering and opening up for viewing what they saw around themselves. Here, the emphasis was not on collecting but on finding beauty in the everyday and then framing it, placing it in a particular way, or putting it in a museum context so that its inner value would come out.

The next step was not just to exhibit the object but to challenge it to bring forth whatever meaning or beauty it contained. The Gutai movement (1954–1972) in Japan delighted in breaking or fragmenting things, as did the Italian painter Lucio Fontana (1899–1968), whose most famous trick was to precisely and with great panache rip an empty canvas and then exhibit the results. However, these remained resolutely art objects you could view by themselves.

In the 1970s, the artist Gordon Matta-Clark—who, not coincidentally, was trained as an architect and was a son and nephew of artists involved in the surrealist movement's attempts to evoke dreamworlds—was one of the first to bring together that critical attitude with the assembled constructions of Soane and Schwitters. He first made his mark in 1974 by taking a Sawzall, a versatile tool for construction and deconstruction, and cutting precisely down the middle of a house in rural New Jersey owned by his gallery dealer. As the result of the surgery, the two halves sank in a lopsided manner away from each other on their foundation. The cut also revealed the insides of the house, making what was private public while turning a large object of use—a residence—and its various parts into a work of art. In this mode, it might not have been a design for living, but the visibility of the structure's various interior and external elements showed architects how they could open up existing structures.

Matta-Clark went on to cut into a host of other buildings. Taking his Sawzall and other implements to a building in Paris that was soon to be demolished next to the Pompidou Center, he created *Conical Intersect* (1975). He cut circular areas out of the structure's layers of walls so that, from the outside, you could see what appeared to be a spiraling void that offered a dizzying view of what had been an apartment building and the street beyond. The effect was a delirious version of Sir John Soane's house: the vistas through the opened-up rooms functioned—like Soane's interior windows—to give skewed and compressed views of the dense records of history present in the structures. *Conical Intersect* was also a commentary on period rooms, in which the wallpaper, sconces, bits of floor, and faded areas, once occupied by furniture and framed paintings, became a continuous composition in space.

In such work, a fusion of architecture and art emerged that took the form of an archaeology of the recent past. It was the reverse of the reconstruction of period rooms but also a way of turning what remained of a structure into a collection: what was left was no longer just walls, floors, ceilings, wallpaper, and other parts of a building, but fragments whose composition—created by the cuts—were part of what made the site now a work of art. These remains offered a radical alternative to what we usually think of as art, but also still identified themselves as actions or works of that nature.

This is the context in which Ireland was reconstructing his home in the 1970s. While parallels to his work were emerging around the world at the same time, I know of no other site that brought together the construction and materiality inherent in human artifacts with the notions of archaeology, selective preservation, and the display of the past. The house at 500 Capp Street drew on many sources, but it coalesced those influences into the new idea that an act of renovation could also be an act of archaeology: a collection of memories in built form and a work of art.

THE INDUSTRIAL ARCHAEOLOGY OF MASS MOCA

We know of one direct influence Ireland's work had. The architect Henry Moss visited 500 Capp Street in the late 1990s when he was preparing to renovate several abandoned textile mills in western Massachusetts into what would become MASS MoCA (Massachusetts Museum of Contemporary Art, founded in 1987 by Thomas Krens and financed by the Commonwealth of Massachusetts). Moss's design, like Ireland's, highlighted the textures of the original buildings, preserved some of the fragments of their industrial past, and composed the remains into artful focal points that at times competed with the works of art on display. MASS MoCA has been hugely successful and influential, drawing millions of visitors to an area that had been an economic backwater.

Moss did not copy Ireland and was certainly also inspired by other sources, but he did pick up on the incompleteness and somewhat fetishistic attention to surface and materiality at 500 Capp Street. When you walk through MASS MoCA's Building 6, the first mill that Moss renovated, you will encounter the white walls and spotlights that are the mark of a modern art museum, but you will also find bits of green and gray paint on the iron columns that the architect found on site. Similarly, the brick walls display some of the many coverings that had been applied to them over the years, while the window frames offer a contrast with their own discordant paint colors (see image #7).

More than that, Moss and Krens accepted the buildings. Each building had been constructed over the years to accommodate different uses and machinery and was thus itself an assemblage. Though each of its parts had once had a purpose, that original context was now gone. The various industrial brick, wood, and steel-framed buildings—the Arnold Print Works and the Sprague Electrical Company being the two main constituent plants—jammed into each other, often at strange angles, leaving slivers of space both inside and out. Moss put skylights over some of those, turning them into indoor atria, while leaving others empty. Where brick of different eras collided or beams intersected as they supported separate constructions, their meeting points became accidental compositions that could often vie with the sculptures shown inside.

Threaded through the industrial archaeology are new stairs, lighting tracks, bridges, corridors, and all the other elements needed to make a modern museum work. The whole complex became a playground for artists whose work extended to performance and site-specific installations.

Visitors came as much to marvel at the old industrial site—opened and made inhabitable, sporting its unused artifacts and displaying the machinery that had made them work—as they did to see works of art framed by all that heritage. This itself became a template with which architects began to construct a model for how they could make their work relevant.

MASS MoCA has spawned a host of similar projects, both in the United States and elsewhere: the DIA Art Foundation in Beacon, New York, located in a former printing plant; the Mattress Factory in Pittsburgh, housed in, as the name indicates, a former mattress factory; a former textile factory turned into De Pont Museum in Tilburg, Netherlands. The most successful imitations turned up in China, starting with 798 Factory in 2000, a former East German munitions plant in the heart of Beijing that was renovated bit by bit into galleries, tea shops, artist's studios, and eventually stores and private art museums. That venture itself spawned imitators across China, including the OCT complex in Shenzhen, the West Bund Art Center in a former airplane factory in Shanghai, and Xiaozhou Art Zone in a group of renovated warehouses in Guangzhou, which includes Redtory Art Park and Taigucang Wharf. Now every self-respecting city has to have a renovated factory—its steel beams and worn-out brick walls fully on display around cutting-edge art—in order to compete with its regional rivals.

By now, the history of artists transforming such buildings has been almost completely submerged by commerce, though the strength of the original buildings is such that they continue to provide an effective counterpoint to the slick surfaces of the new functions and the even slicker objects that serve those activities. Soon stores were catching onto the sensuousness and dreamlike quality of 500 Capp Street. Clothing shops displayed their garments against only partially restored walls, hanging wares directly from the remains of machinery or implements. Displaying items in found containers, like big jars or medical cabinets, became commonplace. In the early 2000s, the Japanese fashion designer Rei Kawakubo turned such displays into replicable models when she opened her first Dover Street Market complex of stores in London, then copied that idea in other cities around the world. By now, what Ireland wrought has become as well-worn as the white walls and skylights that used to denote museums.

That preference for rough walls, however, is also tied to a larger change in how we define value and meaning in the field of clothing. In many ways, Dover Street is based on the aesthetics of the punk movement, which delighted in torn clothing, reused and reassembled rags, and material that was as grungy as possible. The punk revolution and the movements that came out of it made it possible for people to think of their outfits as being a combination of high and low. They could assemble outfits out of T-shirts and brocade, or torn jeans and retro black tailcoats. Collecting these fragments became routine for some: the look kept changing as the wearers collected and discarded elements of their daily uniform.

WORLD OF INTERIORS

In the 1980s, the domestic realm also found itself folded into a style that fed the popularity of stores such as Dover Street and countless imitators. This style was popularized above all else by the English magazine *World of Interiors*, which first appeared in 1981 as *Interiors*. Edited by fashion maven Min Hogg, for the first two decades of its existence it specialized in showing the grand interiors of aristocrats' inherited homes in all their decrepitude and lived-in qualities. Unlike other decorating magazines, it did not shy away from displaying mismatched furniture and bric-a-brac that ranged from the valuable to junk. The magazine accepted and showed it all as it was put together by owners building a life for themselves within the carapaces their forebears had left them.

In a cynical sense, the beautiful photographs of messy spaces made the lives of the illustrious seem more attainable to those aspiring to a grander position, but Hogg also showed humbler interiors in which the inhabitant's skill at selectively preserving and adding collected artifacts—be they dolls or antique sculptures—made up for the rather more modest scale and history of the home. As the 1980s wore on, retro fashion designers such as Todd Oldham and Vivienne Westwood turned that sensibility into a style you could buy—frock coats, Victorian corsets meant to be worn as shirts, and leather jackets festooned with medals and other knickknacks—if not off the rack, at least in the showrooms.

World of Interiors also had a larger effect. It brought the century-long effort by artists to revalue the leftovers of our daily lives and see them as works of art—thus critiquing what we thought was valuable—into the mainstream of our cultural experience. It also made eccentric collecting fashionable. In so doing, *World of Interiors* put front and center the idea of the act of collecting itself as a work of art—a ritual, a continual performance, and a three-dimensional assembly.

The renovation that David Ireland accomplished in the years after 1974—which produced a result that could have been, but never was, featured in *World of Interiors*—accomplished something similar. Although Ireland probably did not set out to do so, he wound up making an argument for architecture that places us in memory and history. Thus, 500 Capp Street both connects you to other people's lives through the material ghosts they have left behind and makes you aware that you are part of a larger history of building.

EVOKING WORLDS

Ireland's strategy was an answer to a very basic problem that architecture was confronting at the time and that still plagues it. Architects strive to create a better world—whether by making a more pleasant home that works more efficiently, constructing social houses as the building blocks for a more perfect society, or designing state institutions and public spaces, and even organizing cities. Time and again architects' efforts have failed, in large part because what they produced was, inevitably, new. What is new is alien, other, and strange by definition. Move into a new house or office and you have to make it your own, bringing your furniture, family photographs, or even wallpaper to cover its newness. Every act you do to make the building familiar hides more of what the architect has done—ironically enough making it even more alien and outside of your daily life because you cannot see the logic of how the building was composed and is put together. It is one of the reasons most people hate modern architecture.

Basically, 500 Capp asked: what if, instead of starting with something new, architects began by figuring out where they were and where they had come from, both individually and as part of a collective? One power of art is its ability to evoke a world that is strangely familiar, to make you realize you are continuing in patterns and quests that others have participated in before you. We turn ourselves into heroes and heroines of the past, we recognize ourselves in portraits of people long dead, and we follow in paths worn by communal use.

With buildings that are a combination of archaeology and collection, architecture can do the same thing. It can reveal the work that went into making its beams and columns, its frames and floors, but it can also define the different ways in which previous occupants once made that place their own. The traces of wallpaper or paint are not only pretty in themselves: they are also evidence of choices people made, and the tastes that went into their decisions. As we scrape away at buildings to make them show what they had been, we reveal that, while these structures might be just generic edifices put together according to general standards, the people who built them had left their mark on them and generations who lived there had layered on their own patterns.

This strategy breaks through the hierarchies that usually establish what is important or valuable and what isn't (is a broom as beautiful and expensive as a Picasso?) and thus offers an alternative to the social and economic structures that define not only art but every aspect of our lives. The renovation of 500 Capp Street inspired imaginative reuse but also provided a trenchant critique of the society in which it appeared at the end of the twentieth century and which remains relevant today.

CHAPTER 5

SQUATTING, INSTALLING, AND ACTIVATING

THE DORCHESTER ART PROJECT

I walked down Dorchester Avenue on Chicago's South Side and knocked on the door of the house with no paint. Slats of wood, mismatched and bearing marks of previous use, made up the face of the building. I did not have an appointment here, but I had decided to try my luck. The door opened and a head popped out. "Hi," I said. "I am here to see the Dorchester Art Project."

"And you are?" he asked.

"Aaron Betsky," I answered.

"Of course you are. One moment." I guessed that he'd recognized my name from the criticism I write. The door slammed shut. I stood around for a few minutes, admiring the different grains and marks on the wood. When the door opened again, the man had covered his cheeks and forehead in dabs of color that he later told me were intended as war paint. "I am ready now," he said, letting me into the house and launching into a nonstop half-hour diatribe about the ways in which architecture only served to perpetuate capitalism and how we could escape from that condition by refusing to play by its rules, choosing instead to work with communities on reconstructing their places and structures. Pamphlets and manifestos pinned up on the bare studs of the interior explained his often-breathless words (image #8 shows what I found). "Any questions?" he asked when he was seemingly done. I had many.

The man turned out to be Xavier Wrona, a French architect and professor who was then in residence (I visited in 2015) at the Dorchester House Project, an art project, residency site, and community activity hub. Established by the Chicago-born Black artist Theaster Gates when he bought the abandoned home in 2009, the project is a recycling of itself: Gates stripped the structure down to the bones, then used the scraps not only to reconstruct the structure's covering but also to build furniture. He added building material he accumulated in his workshop and also soul and R&B records from a defunct shop that had once been a focus of this historically Black neighborhood. Old issues of *Ebony* magazine joined the collection, and all of this went on display in the house. The wood studs and bare window frames speak to the absence of the tidy domestic life that once was led there on Dorchester Avenue. The original inhabitants are long gone, but the bones of their home—reassembled into a fabric that merges table into wall into floor into chair into cabinet, a web of worn wood redolent of use and decay—not only evokes their existence but continues it in the speculations of the artists, philosophers, and architects who work there.

The site is also, as Wrona had shown me, a site of performance, which is central to Gates's work. There is a theatricality to his acts of deconstruction, reconstruction, reassembly, and repurposing of structures. He also gives performances there with a group of musicians called the Black Monks.

The Monks play in a style that combines blues, gospel, chants that might be Buddhist or Tibetan, and other shards of melodies, rhythms, and motifs. Over the course of a performance, the music may start out meditative, developing an aura that pulses around the unfinished spaces and wraps around the exposed rafters. Chants rise, a beat appears, and soon the small troupe is in full swing, throwing out parts of recognizable phrases ("there is a house in New Orleans . . . ") and incantations that are more difficult to decipher. The audience might join in, rocking or even dancing, shouting their approval or singing along. The performance sings alive the sites where it occurs, suggesting the place has become a funky chapel, a site where people are dedicated to knowing, or at least sensing, the meaning of the place. I have seen the Monks perform in an abandoned apartment building in Kassel, Germany, which Gates had turned into an art piece by partially renovating it and using the demolished materials as stand-alone art pieces. I have heard them on the porch of a house in

Napa Valley as night fell over the vineyards. And I came upon them in New York's New Museum one Sunday, when the elevator opened and Gates faced me directly, surrounded by his fellow musicians, belting out something I could not understand but that drew me in.

HUNTER-GATHERERS, MAGICIANS, AND BRICOLEURS

The Black Monks are part of the places Gates constructs. He repurposes familiar forms, places, and sounds to evoke something that we otherwise have lost or just have not yet found. He is an assembler, as he himself says, "gathering, stripping" material to emphasize that act itself: "Stacking, restacking, and shifting. Piled lath, piled old-growth piles, potent and latent, piled histories, accumulations, and other such notions. Neatly and sometimes not so neatly, the gathered things start to suggest forms." As a result, he writes, "We see the forms, and our need (ambition) sometimes determines what happens with the pile." What he winds up with are piles and stacks but also assemblies that are the "result of lots of hands and hammers, pull bars and moving straps, time spent thinking, dreaming, and sorting."

"We are the hunter-gatherers," Gates concludes, "ever finding the accumulated, the forgotten—the oh so stackable."[1]

With this self-definition as a "hunter-gatherer," Gates places himself in a particular tradition that offers an alternative to the standard notions (at least in the West) of what art and architecture are. It was articulated in the modern age most clearly by Claude Levi-Strauss in *The Savage Mind*, which defined the methods of the "hunter-gatherer"—a person who depends on both hunting animals and gathering plant material for survival—and distinguished between the realms of "science" and "magic." For him these terms corresponded, respectively, to the Western modes of analyzing the world and, in the demeaning language of his time, the "primitive" world.

Levi-Strauss saw science as a preformed scheme that tries to fit what it finds within its interpretative and statistical categories. It makes sense by comparing, creating hierarchies, and positing abstract meaning out of the relations between objects. It seeks to create something new and previously unknown, which it assumes is better. Magicians, on the other hand, gather together the elements around them to find patterns, relations, and meanings that are already there and that they only affirm through their arrangements. They seek not to make something new but rather to restate, rearrange, and gather together the meaning that is already there.

Levi-Strauss calls the person who works in this manner not just a magician but also a "bricoleur," somebody who collects cast-off materials and assembles them into a collage. That is exactly the opposite of a scientific artifact: it is nonhierarchical, not new, and not abstract. In fact, as an assemblage, it does not appear to be ever quite finished. The bricoleur, or collage artist, delights in the different textures, forms, and even images they find and leaves them as they are. By bringing them together in stacks or piles, as Gates does, they produce other relations, perhaps even overtones like the ones Gregorian monks might make when they chant.

Theaster Gates came to this approach from a varied background. He was trained as an artist, an urban designer, and a theologist. He also apprenticed with a master ceramicist in Japan. His father was a roofer, and Gates has used his father's craft and tools in several of his pieces. At the time he came to Dorchester Avenue, he was working as a community activist with South Side neighborhoods and the University of Chicago. That combination of craft and theory, a search for meaning and social sense, and a profound commitment to the Black community, its culture, and its history, all led him to his act of dis- and reassembly.

Since he purchased and transformed the first house on Dorchester Avenue, Gates has added several nearby properties. A former bank building around the corner on Stoney Island Avenue has become the headquarters for his activities. There, as well, he stripped and stacked, piled and revealed elements of the existing building to turn it into a combination office, archive, studio, and community center.

Gates has gone on to carry out such hunting and gathering in other locations, including at international art fairs and across the American Midwest. To keep his operation going, he takes fragments of these (de) constructions and sells them as artworks, commanding hundreds of thousands of dollars for many of them. That is not so strange as it might seem: he is working in a recognizable collage and assemblage tradition that the art world values (in both senses of that word), and he is quite good at what he does.

Such work has made Gates the leading avatar of a mode of making that exists at the intersection of art, architecture, and community action. The movement combines two traditions: site-specific artworks and squatting in unused buildings. By bringing in the squatting tradition, it takes the work of artists such as Schwitters and Ireland one step further into the social realm.

SQUATTING AS DIY REUSE

Squatting is the most common form of do-it-yourself reuse you can find in architecture—not just in the West and since the emergence of the squatting movement in northern Europe in the 1960s, but throughout the world and the ages. People have always used abandoned structures—whether they were the ruins of Roman civilization or former fortifications—and turned them into apartments and villages. Squatters made these structures their own, using what they found to create interiors that reflect and frame their particular lives.

What was different in the squatting movement of the 1960s was the self-consciousness and deliberateness of the occupations. It started in Denmark, West Germany, and the Netherlands, all of which were suffering through a housing crisis even as many properties remained vacant, either because landlords sought to maximize their investment or because the structures had been built for nonresidential uses. Groups of young students and unemployed people, often allied with anarchist action groups, began to take over these structures, barricading themselves inside and developing social systems for communal cooking and eating, childcare, and sanitation. They scrounged discarded furniture off the streets or made it themselves out of recycled material.

The movement soon expanded across Europe, with the largest squat becoming Christiania, a "freetown" of between eight hundred and a thousand people located on almost twenty acres of a former army base in Copenhagen (see image #9). Founded in 1971 and still in operation, this commune developed not only more sophisticated organizational structures that proved capable of negotiating and renegotiating its existence with local authorities over time but also a distinct aesthetic. Rooted in early twentieth-century art movements such as Arts and Crafts and art nouveau—which many of those squatters had learned about from their training as artists or their reading about utopian historical communities—this hippie aesthetic merged with the rougher look of the spray-painted slogans and images these activist squatters were producing. The decor also delighted in its own sense of incompleteness, rejecting the "proper" arrangement, treatment, and finish of walls and furniture as emblems of a hierarchical, middle-class way of life.

Rooms in the middle-class home since at least the Victorian era had been arranged around certain orders and rituals: separate bedrooms for parents, children, and servants, bathrooms with privacy, kitchens segregated for either servants or women of the household to cook, and areas designated for gathering. Within these spaces, furniture designated which person should use what chair, table, or implement: there were pieces intended to be used only by men, only by women, or only by children, for instance, with their size and design supposedly tuned to different body types, but also to what society considered the correct aesthetic for a certain gender or age. Together, the spaces and their furnishings all made the social order into lived facts.

The squatter movement reused objects in the "wrong" way, turning sofas into beds, putting mattresses on the floor, cobbling tables together from sawhorses and doors, doing away with privacy divisions, and using spaces for whatever they seemed to be best suited for at the time. The decayed nature of the old buildings was usually left as it was, except for absolutely necessary repairs to plumbing and electricity. Even those were often not pursued, not only because there was no reason to fix up buildings from which you might be evicted at a moment's notice but also because that was a sign of the powerlessness of the organizations that had originally created those walls, wallpaper, and sequences of spaces.

America has had its own squatters, with activists taking over buildings in New York City, Cambridge, Massachusetts, and Berkeley, California (where they also famously squatted in what was meant to be a building site for a new dorm, turning it into People's Park in 1969), but here the movement started later than in Europe. Although there were some informal and small-scale squats in the 1970s, both apartment buildings and single-family homes here were protected by strong ownership laws, backed by the authorities. This stood in contrast to the more liberal (or social democratic) policies in Europe. Only in the late 1970s and early 1980s, when major cities on the East Coast faced a severe economic crisis and the wholesale abandonment of housing stock, did larger squats come into being.

These squatters were, as one observer has noted, more punk than hippie, with little regard for the physical setting in which the squatting took place, and also were less communal in their character.[2] While one building, C-Squat in Manhattan's Alphabet City, did transform the shared stair hall into a communal space, most of the others remained collections of individual living areas, as they had been when they were apartments.

These New York squats were legalized in the early 2000s, and only C-Squat has retained some sense of its history—albeit a rather restricted one—by converting its storefront into the Museum of Reclaimed Urban Space, which preserves and displays graffiti and found artifacts from the building's history. Those artifacts and the way they were collected recalls what Ireland had done before, and Gates later, in their art installations:

> Sometimes the history is other people's and is literally in the building—a squatter is demolishing the last scraps of plaster walls before putting in new drywall or knocking out an old chimney and finds some clue to the past: a book printed in Hebrew, the edges scalloped by mice, a wooden spool, or a handmade screwdriver, still useful in the present. While some probably passed these by, adding them to the rubble bucket to be dumped with the broken bricks and plaster in a vacant lot, others treasure them.[3]

As Amy Starecheski, who has documented and analyzed the New York movement, notes of the squatted structures: "[The] buildings, many on the verge of collapse when they were occupied, were preserved through their efforts and will now shelter the aging squatters and their children, a small bastion of affordable housing in a heavily gentrified area."[4] These former squats, by their actions in making the buildings safer, might have contributed to the improvements in their neighborhood. It is only in "the private, secret, and hidden spaces of the buildings," Starecheski notes, perhaps overly optimistically, that "they preserve their stories by planting artifacts that will eventually grow into history for a new generation, perhaps several generations into the future. Through these history-making practices, they mobilize their connections with those who lived in their buildings nearly a century and a half ago, who perhaps also marched, chanting, through Tompkins Square Park, with present struggles, and with an imagined future." (Image #10 shows one of the apartments.)

Together, the aesthetic and the activism of squatting created an alternative for urban living. It gave value to unused buildings, not only for their availability but also for the freedom that they implied. Without the constraints of family or class arrangements, new forms of sociality could emerge, allowing squatters to construct their own new world out of time and place.

The squatting movement lasted only so long before it was largely destroyed—squatters were evicted from their sites or, alternatively, their living arrangements became regularized and thus normalized by authorities, as in the fairly elaborate modes of self-government Christiania has created. Their aesthetic now looks like a relic of the particular time in the 1970s, when they were at their peak.

The squatting that has remained as an active movement is smaller scale: occasionally people will still take over buildings in Amsterdam or Berlin, but it is very rare. It has turned into a form of art, holding onto its activism even as some of its actions are now legal. In the United States, there is very little squatting in the inner cities, although anecdotal hints of squatting in failed suburban developments certainly deserve investigation.

Activist reuse has usually involved, as in the Dorchester Project, buying the property. The most famous and enduring such invasion through purchase was and is Project Row Houses, a complex of small houses in Houston's historically Black Third Ward. It started in 1993 when the artist Rick Lowe purchased a small house there and began using it as a studio and community center. Today the organization encompasses thirty-nine structures over five blocks and is one of the city's main engines of activism and education through art.

THE HEIDELBERG PROJECT: RIOTS OF COLOR AND FORM

The Heidelberg Project in Detroit's McDougall-Hunt neighborhood (named, like the Dorchester Street Project, after the street where it is located) is of a completely different order. I discovered the project years ago when I went looking for the street where my mother grew up. What I found (see image #11) was a largely empty block, sporting only a few forlorn and boarded-up structures amid trash and resurgent nature. Driving off, my husband and I passed through a landscape of devastation that stretched for what seemed like miles—the result of the decades-long collapse of Detroit's economy and the abandonment of many of its inner-city neighborhoods—until we came across a sudden organization of the trash into piles. We stopped and realized that these piles actually contained objects of use that had been organized, albeit loosely, with some purpose. Discarded clothes and shoes made memorial sculptures to those lost.

Shopping carts became both pedestals and containers for riots of color and form. Polka dots were everywhere—on concrete walls, the discarded objects, and houses. Though I did not see many people other than the volunteers working at the Project, it seemed to me that all the life of this former community in what was once the nation's fourth largest city had been sucked into this one block, organized according to principles that were difficult to figure out, but evident and highlighted with paint. All of this, I would later learn, was Tyree Guyton's Heidelberg Project. The critic Marion E. Jackson described the scene as she found it: "Large painted dots . . . stuffed animals squeezed into boarded window frames . . . shoes swinging freely from golden branches of autumn trees." She also saw "a parade of swaggering vacuum cleaners . . . boldly painted faces smiling from bent car hoods . . . star-spangled banners wrapped around trees and painted on weathered boards . . . bedsprings . . . posters . . . street signs . . . long-lost dolls with missing limbs and dented faces . . . shoes . . . bathtubs . . . wooden crosses . . . hubcaps . . . and dots . . . colored dots . . . more dots . . . " and then "an entire house painted with polka dots . . . a photograph of Martin Luther King Jr . . . and dots, more polka dots."[5]

This continuous (in both time and space) work of art began in 1986, when Guyton, thirty-one, was being trained to paint by his grandfather, Sam Mackey, who was also a resident in the then almost exclusively Black McDougall-Hunt community. One day, as he recalls, Guyton looked out from Mackey's porch and began envisioning the already largely empty area as his extended canvas. Like a bricoleur or hunter-gatherer, he set about using what was at hand—including not only what he found on the street, abandoned or unused but also his paint and paintbrushes. He used the latter to create patterns that did not so much unify the various objects and spaces he found as draw relations between the disparate elements while evoking a half-Christian, half-abstract, and all-Black pop culture iconography, with everything from crucifixes to photographs of Black cultural icons.

Guyton describes his work as nothing more or less than "composition," which he also sees as a way of "traveling into the unknown of my unconscious mind."[6] By rejecting the hierarchies not only of what is high and low art but also of what is made and found, what is pattern and what is form, and, quite simply, what is put on a pedestal and what appears to be strewn in a yard, he rejects the "traditional and conventional way" to "tell a story—my story, our story—about life and what I see in the world." Other critics have compared his work to that of Eshu, the trickster god in Yoruba religion who transforms the normal into the divine and vice versa, presenting realities that can be either true or false, ugly or beautiful.

Over the years, Guyton has bought or simply used several other abandoned houses, yards, and vacant spaces around his grandfather's neighborhood. He has also developed art classes for local children on the site, seeking to duplicate how he came into his vocation, and he has formed a community center. He has lost some of the art to a city government that is ambivalent about his work, at times supporting it with grant money and publicity, and at other times moving in with bulldozers to tear down structures they felt were unsafe or unsightly under ordinances developed to combat blight in the area. Neighbors are similarly both supportive and antagonistic, in the latter case not only because of art that might be trash but also because of the steady stream of visitors such as myself who come to see the project, taking pictures and gawking.

The Heidelberg Project—which Guyton has by now largely abandoned to move onto new projects, leaving behind a foundation that maintains the space—is indeed a trickster site. It is both art and architecture, old and new, object and field. It is rehabilitating and giving life to a site that had lost most of its social structure and hope, but it might also be a form of gentrification—although there is little sign yet of the latter in the neighborhood. In that sense, Guyton's work points to a larger dilemma that Detroit is facing. Having reached a nadir during the second decade of the twenty-first century, when it was severely underpopulated because of the collapse of the local car manufacturing industry, it is now revitalizing itself, largely by reusing existing structures. Open fields such as those around the Heidelberg Project have become sites for other art projects but also for urban agriculture. Similarly, other artists and collectives are streaming into the abandoned factories and homes to turn them into places of both production and display, albeit for a limited audience of art aficionados. Meanwhile, photographers such as Camilo Jose Vergara have used the abandoned buildings to create artfully composed shots saturated with color, now often called "ruin porn": seductive images of decay that you can hang over your sofa. They are beautiful but raise the question of whether we should be enjoying sites of economic and social despair for our own aesthetic pleasure.

1. Reused Roman columns in Santa Maria in Trastevere ©DAVID MERRETT

2. "Do" Design: a cube of metal can become a seat ©DO HIT, MARIJIN VAN DER POL FOR DROOG DESIGN. PHOTOGRAPHY BY JOKE ROBAARD

3. Simon Rodia's Watts Towers in Los Angeles ©MOE.217

4. The Grand Shrine of Ise, Japan ©TETSUDOIN

5. Instant bridges and communities on the Ganges River ©BRETT COLE

6. The interior of David Ireland's house at 500 Capp Street, San Francisco ©WIKICOMMONS

7. Building 6, a renovated factory building at MASS MoCA, North Adams, Massachusetts ©DOUGLAS MASON

8. Interior of Theaster Gates's Dorchester Project, with installation by architect Xavier Wrona ©AARON BETSKY

9. A street in Christiania, Denmark ©SCHORLE

10. One of the apartments in C-Squat, New York ©KONSTANTIN SERGEYEV

11. Tyree Guyton's paintings and assemblages at the Heidelberg Project, Detroit ©AARON BETSKY

12. Working out on a high floor of the unfinished Torre David, Caracas, Venezuela, during its occupation ©DANIEL SCHWARTZ / CHAIR OF ARCHITECTURE AND URBAN DESIGN, ETHZ

13. A loft renovation in New York designed by Alan Buchsbaum reflects the aesthetic of such spaces in the 1970s. ©BILL MARRIS AND JULIE SEMEL COLLECTION, PRATT SCHOOL OF INFORMATION ARCHIVE, CREATIVE COMMONS

14. Spanish architect Carles Oliver's renovation of an old home into new apartments ©DANIEL SCHWARTZ / CHAIR OF ARCHITECTURE AND URBAN DESIGN, ETHZ

15. The Winter Garden at Granby Four Streets, Glasgow, Scotland, designed by Assemble Studio, London ©AARON BETSKY

16. Carlo Scarpa's renovation of the Castelvecchio, Verona, Italy ©AARON BETSKY

17. Powerhouse Arts, a former power station redesigned by architects Herzog & de Meuron ©ALBERT VECERKA/ESTO

18. New York's High Line USDA PHOTO BY LANCE CHEUNG. HTTPS://CREATIVECOMMONS.ORG/LICENSES/BY/2.0/

19. Artists can provide us with an epic version of imaginative reuse by restaging a worn-out reality and infusing it with religious and redemptive overtones. GREGORY CREWDSON, *REDEMPTION CENTER*, 2018-2019, DIGITAL PIGMENT PRINT, IMAGE SIZE: 50 X 88.9 IN. © GREGORY CREWDSON.

20. The LocHal library and community center, a former tram repair shop renovated by Civic Architects ©AARON BETSKY

21. Photographer Manuel Alvarez Bravo's former home in Mexico City, Mexico, now a hotel designed by Ambrosi Etchegaray ©AARON BETSKY

22. A swimming pool in the Zollverein, a former steel plant in Essen, Germany ©AARON BETSKY

23. The 798 Factory in Beijing, China, uses a former arms factory as its site. ©CHARLIE FONG

24. The central hall in Brooklyn, New York, from which the US Army shipped out goods around the world, is now becoming an incubator for new businesses. ©AARON BETSKY

25. "Waiting Room," an image from Wanda Spahl and Dominic Schwab's project, *Geography of Ghosts* ©DOMINIC SCHWAB

The question raised by these developments is whether the very act of turning urban decay into art (or architecture) takes those sites away from the inhabitants who once made them or still use them. Is this kind of re-imagination a form of appropriation akin to what high art has done with other "outsider" cultures? And, more importantly, does it do so as part of an act of gentrification in which not only the artifacts but the very homes of those who still live on these sites are taken away from them? Does it matter if the work is being done by artists who are part of the community, or does their artistic use of the buildings simply remove them from housing stock that should serve that community? These questions cannot be immediately answered, especially not by outsiders such as myself.

The area around the Heidelberg Project shows only tentative signs of "upscaling," while most of the urban agriculture and community art projects are, if not community-led, certainly very careful to be part of the places and the social networks in which they are situated. For now, then, there is little proof that the art of reimagination by itself causes gentrification, mostly because such art is a small factor within much larger economic and social forces. A neighborhood like San Francisco's Mission has become largely gentrified, but that is because of the immense pressure on housing in the limited space of the Bay Area.

TACTICAL URBANISM

The tactical urbanism movement raises similar questions. The movement has its roots in the acts of queer men (and some women) who, starting in the 1940s and well into the 1980s, cruised abandoned buildings, wharfs, and parks in all major Western cities to find partners. Their activities brought life to these areas and, over time, attracted other forms of activity. The most well-documented queering of unused space took place along and in Manhattan piers where, starting in the late 1960s, a cruising community extended its search for sex to sunbathing, dancing, or just gathering, attracting bars and other commercial activities to the surrounding streets. Though the cruisers sometimes left marks and graffiti, it was rather the appreciation of such spaces that helped integrate the aesthetic of these decaying industrial buildings into the culture of which many of these LGBTQ people were part. Artists followed—most notably, as we've seen in chapter 4, Gordon Matta-Clark, who in 1975 transformed one of the warehouses there through his work *Day's End*. This was itself an installation (or anti-installation, as it consisted of a cut through one of the walls in the shape of a tilted lozenge) that recalled the acts of revealing and recuperating of memory in which David Ireland and Theaster Gates engaged.

Tactical urbanism began in earnest in South America in the 1990s, as both architects and activists sought ways to take charge of underutilized sites and buildings. Their efforts built on the history of graffiti, which marked buildings with human presence, and on the squatting movement. Groups such as the Venezuelan architecture firm Urban Think Tank began by selectively taking over parts of a city and turning them into architecture that is also art. Such approaches have blossomed across the world in a wide variety. In Tirana, Albania, the mayor authorized painting whole parts of the city in bright colors and patterns to make it more attractive. In the United States, people began gardening in empty lots in the Bronx and Detroit. Other interventions have ranged from weaving patterns into chain-link fences with everything from more fencing material to macramé ("knit bombing") to Victor Solomon's 2020 tracing of the cracks of a basketball court in Los Angeles with gold leaf.

Yet most of these works remain adornments of existing buildings, usually abandoned or underutilized. They might make them visible and even more active parts of a neighborhood, but they do not have quite the sense of creating community through the re-inhabitation and singing alive of existing buildings that artists such as Gates and Guyton were able to achieve. That is work that involves a larger effort, one that often skirts between legality and illegality, between permanence and the ephemeral.

TORRE DAVID

The most notorious action in recent years that has re-inhabited an existing structure was the 2007 invasion of the Torre David, an unfinished, forty-five-story office tower in Caracas, Venezuela. The tallest building in the country, the complex had sat empty and half-clad since 1993, when its financing evaporated and its namesake developer, David Brillembourg, passed away. The invasion began with a group of local homeless people, organized by community activists, moving into the structure, and occupying it all the way up the twenty-fifth floor. They set about making themselves at home, converting stretches of concrete office floors into homes separated from each other first with sheets and tarps but soon enough also with walls of brick and concrete block, sometimes covered with wallpaper. Community spaces also appeared: a church attracted a large congregation, a basketball court was marked off in the courtyard below, and leftover spaces were outfitted to encourage people to gather. Commerce appeared as well: a barber shop opened, as did several small shops. You could get your cellphone fixed and obtain legal advice within the rabbit warren of the Torre David. Torre David the building was also featured in the American series *Homeland*, serving as a site where a renegade soldier, who has also been a politician and an Iranian spy, hides out.

What was remarkable about this latter-day vertical squat was its informality and lightness of touch. The occupants organized themselves to manage communal issues such as access, water and power distribution, and waste disposal, but they never really developed a government that told people what to do. Disputes about territory were few, as the building was vast. That also meant that a certain amount of illegality crept in, and in 2012 police invaded the complex looking for a drug dealer.

Almost as soon as Torre David became occupied, architects and advisors showed up, most notably the Caracas firm Urban Think Tank mentioned above, whose practice focuses on developing infrastructure and community buildings for informal settlements (ironically, one of its two partners at the time, Alfredo Brillembourg, was a nephew of the tower's original developer). Over time, the firm not only documented the spaces but began proposing ways in which a more permanent settlement could emerge. They suggested sustainable energy systems, new circulation elements, and facade treatments that would both protect the inhabitants and create a new image for the Torre David.

What they did most of all, however, was to listen and look. Working with the Dutch photographer Iwan Baan, they produced images of life and architecture in the building (an example by a different photographer is image #12), disseminating them in books and exhibitions that made the invasion famous around the world. In 2012 they even won the Golden Lion, the highest award for a project, at the Venice International Architecture Biennale.

What made the photographs so powerful was the beauty they revealed in the lives made within this abandoned building. Some of Baan's images emphasized the drama of the situation: one shows an inhabitant bench-pressing weights on the empty expanse of one of the unfinished upper-level floors, positioning himself and his equipment right next to the floor's edge, the landscape of Caracas spread out below their pumping arms. The empty space and the lack of barriers create a sense of vertigo, making the building look more heroic and grander than it ever would have been as a finished structure. But Baan also showed human beings occupying what seemed to be crevices, perches, and cantilevers, shifting your sense of scale as you look. The stairwells without stairs, the openness of the facade, the shacks and stores fashioned inside of spaces originally meant for big business, and the kids playing underneath towering forms all showed how Torre David had become an incubator for different relations between individuals and buildings.

If the Torre David community showed that squatting could still occur, it also showed the limits of the movement. In 2014 the increasingly authoritarian Venezuelan regime that had once supported such informal settlements sent the police into the complex, clearing out the inhabitants in a matter of hours. Now the tower has reverted to the way the invaders found it: empty, useless, and decaying.

The message of many of these forms of artful squatting is that they are almost always temporary. They will suffer one of three fates: they will be cleared out once the power of the government or the logic of economic value (developers reclaiming formerly worthless buildings) reassert themselves; they will become works of art like the Heidelberg Project, albeit almost always still under threat; or they themselves become sites, such as C-Squat's Museum of Reclaimed Urban Space, whose economic activity is of high enough value that they become permanent and regularized.

The last possibility is by far the most uncommon outcome, occurring usually in spaces that are barely buildings or even open spaces. In the 1980s, when the Dutch government built a huge new container port in Rotterdam out into the North Sea, the byproduct of that construction was a five-mile beach that the planners built to break up the waves in front of a seawall protecting the port. Instead of planning uses for this stretch of sand, the city waited and watched as windsurfers, nude bathers, party goers, and other users discovered and staked out claims to parts of the expanse. Starting in 2010, and after more than ten years of occupation, the city regularized the uses, giving the different occupants permission to stay there, adding infrastructure where appropriate, and sanctioning the construction of public artworks.

What some governments have realized, in other words, is that letting people occupy or reoccupy space can be a more efficient form of redevelopment than a top-down approach. In such cases, it is artists and activists—who are often the same people—who take the lead.

ASIAT

A recent example of this approach is the sanctioned invasion in 2020 of ASIAT a former army barracks at the edge of Brussels. Like many such facilities around the world, it was abandoned by the military—not because warfare is disappearing, but because the nature of standing armies is, like almost everything else in our economic structure, shrinking and becoming more mobile. In 2018 the local town of Vilvoorde, which had gained control of the half-million-square-foot site, decided not to have developers raise the ruins but rather to let local architects 51N4E work with local activists and artists to open the space for use.

Walk into the area now and you enter into a narrow walkway between the former central administration buildings. The various users and occupants who now populate the site—some of whom pay rent, others of whom just seem to show up—have covered the walkway with a scaffolding of steel tubes and tie-dyed sheets that shelters it during inclement weather. Peek into the windows, which only at times have actual glass in them, and you see a variety of uses and degrees of finish. A ceramics workshop and a gallery have fully outfitted their spaces. Artists are tinkering with work in other rooms with no ceilings. At the end of the walkway, an open-air restaurant occupies a space surrounded by bare brick walls, with remnants of plaster still clinging to them and empty window frames.

Deeper into the territory, nature has taken over what were once parade grounds and neat lawns. This is not an uncontrolled action, though; as one of the architects says: “It is hard work to re-wild.” Working with plant specialists and landscape architects, 51N4E has brought back native vegetation, resisted invasive species, and created moments of density and openness without seeming to design anything.

Works of art are strewn throughout the park that ASIAT now is. Some are traditional objects in space, but most are containers or shelters of some sort. A dome, covered with pieces of plywood, overlapping and somewhat misaligned, defines a gathering space around an open oculus. Similarly repurposed lumber provides the structure. A less formal space is defined by metal tube scaffolding, across which large sheets of cloth are draped. The cloth is hung in artful folds, which not only attract the eye to admire their allusions to Renaissance sculptures but also define a chapel-like space.

The largest of the structures in ASIAT's semi-garden also consists of metal bars, though in this case they are not from scaffolding but rather found metal posts. They shape a rectangular three-dimensional space whose floor is again made of repurposed plywood. A stepped circular depression at the far end invites gathering or, during concerts, indicates where a stage appears. The whole is covered by a collection of plastic sheets—mainly white, but some black or colored—that bear the scuff marks, writings, tears, and abrasions from their previous uses. As in the dome, they do not form a perfect covering, but something more like a general indication and provision of shelter.

My favorite structure at ASIAT is a colonnade that marches between a row of barracks—one of which was being outfitted with a climbing wall when I visited in 2023, while another was already in use as a coffee roastery. These columns are not made out of marble but rather consist of sewage and drainage pipes, precast elements used to support bridges, and some components that I could not identify. Mismatched but of a similar form and finish, they evoke the splendor of some past civilization that might have existed here or the beginning of the (re)construction of another one.

Everywhere you turn at ASIAT there are experiments, constructions, shelters, and activities taking place in and around these structures. Concerts and art classes are held here, and visitors with baby strollers and dogs walk through the park, enjoying the scenery. Economic activity mixes with leisure. The area has been so successful that it is feeling the pressures of overuse on some weekends, but that success has also prompted the energy company across the canal to ask 51N4E to extend their work to some of their former storage areas. These areas sit in the shadow of giant cooling towers, once used for coal and oil, now obviated by the plant's transition to gas.

ASIAT will not last forever, at least not in its current shape. Given the urban growth that is taking place on this edge of Brussels, the land is too valuable, and the current activities have made it more so. Still, there is now a constituency of users and visitors who, the organizers hope, will resist the construction of new building anywhere other than at the terrain's periphery.

Older examples of such transformed spaces—such as the Headlands Center for the Arts, north of San Francisco, established in 1982—have become formalized as art centers, charging admission, inviting artists in residence, and otherwise controlling activity according to the rules and regulations that govern the art world. What sets ASIAT apart is its wildness, its incompletion, and the unstable relation among what is old and remaining, what is new and temporary, what is new and might stay, and a nature that is equally wild and planned.

Projects from Dorchester to ASIAT thrive on such designed ambivalence. They are out of time, in the sense that you cannot determine where their elements came from, how long they have been there, or how long they will remain. They are out of place in the sense that they are assemblages that intensify but do not necessarily make sense of, or build on, their surroundings. They have unclear functions, as they are by turns used as places to see art, gather, activate, and sometimes even live. They change over time, so it is difficult to define, finally, what they are. Are they art, architecture, or temporary structures? It is exactly this ambiguity, this slipperiness, that makes them so attractive and productive. They are the research and development laboratories for the future of communities that might actually be the past, or that might be an alternative reality in which we can all squat, make, and be together, instead of renting or owning, working or recreating.

PART 3

USES

CHAPTER 6

HOUSING: RE-INHABITATION

As it becomes more difficult to create new buildings—both because of land availability and environmental considerations—it is natural to turn to existing structures for that most basic of human needs: housing. Moreover, such existing buildings by their very lived-in nature offer a sense of place and community. The newly built is always other and strange, and that is especially true of residences. We expect our workspaces to be part of an efficient, ever-modernizing machine, and our cultural monuments by virtue of their size and function fall outside our daily experience. But we want to make ourselves at home in the places where we live, and existing structures can do that better than new-built ones.

Repurposing existing structures for housing is a common practice for the simple reason that we cannot build new structures fast or affordably enough. Nor should we try to—there is a stock that we have built up over the centuries. People have made their homes in warehouses, churches, palaces, or office buildings where they could or had to. And as the need for housing increases, mobile work technology has caused the demand for workspace to decrease. We want more space, both inside our living environments and around them, and we want to be as flexible as possible. We also want the buildings we inhabit to make us feel at home, which above all means connecting us to a place and a community. That, in turns, means that these spaces need to evidence and delight in what is already there.

Unfortunately, that is currently generally not the case. Almost all housing renovated from existing buildings—whether they are old residences, former factories, warehouses, or offices—is mindless in its design. They fulfill their function, more or less, but they give little sense of their history or location. The one tradition that allows that sense of place is, ironically, the most basic one: the loft.

LOFT LIVING

As I have written earlier in this book, the loft is the most basic and evocative building block for using, inhabiting, and drawing meaning from what remains of the old. As the functions of loft changed from eighteenth-century industrial workspaces to twentieth-century artist studios, artists began sleeping in the corners of their makeshift ateliers, then building out accommodations within the raw spaces. As these lofts became more elaborate, but also more comfortable, people who visited started imagining themselves living there. Soon "loft living" had taken over much of downtown New York before spreading throughout the world.

The beauty of loft living is that it is both minimal and old. It has all the roughness, worn qualities, and reminders of the age of industry—freight elevators, exposed pipes, perhaps even fragments of equipment—but it also has the kind of foursquare space, clear structure, large windows or skylights, and open layout that we associate with a modern lifestyle.

The loft is cellular, meaning it is usually part of a larger structure. While the ruins of manor houses or workshops will do for single-family homes, factories or warehouses converted into lofts can provide much denser and more flexible housing for different social configurations. In fact, loft conversions have become the basic ways in which downtown areas around the world have been re-inhabited and turned from places of production into places to live and work. Whole neighborhoods—ranging from Hamburg's Harbor City to the warehouse district east of downtown Los Angeles—have become communities made up of such buildings, with the ground floors occupied by restaurants, boutiques, and other shops.

Most of these conversions are rather simple and repetitive and have been documented in countless books and lifestyle magazines (image #13 shows an example). What is more interesting is the way the residential loft has influenced how other structures are now converted into housing. Office buildings can become lofts, it turns out, and so can older apartment buildings. The latter is especially important as municipalities try to figure out how to adapt mid-century, mass-produced housing to accommodate modern ways of living, while also honoring what was already built.

Loft style has certain key characteristics. The first is its emphasis on leaving visible as much of the original work-focused structure as possible. Although by now many people want the vague feeling of a loft without its authentic qualities, lofts are by their nature bare and rough. The concrete, wood, and steel structures are visible in a manner that gives you a clear sense of how the container in which you are living was made. The beauty of bolts and rivets, the grandeur of trusses that come together out of smaller pieces to create a large structure, the way concrete posts flange out to accept heavy loads, and the manner that wood and metal panels are attached to walls such that you can understand how that was done—all are essential parts of the loft look.

The emphasis on daylight is another important feature. Instead of windows being treated as periodic openings in walls, whole stretches of a room are often covered with a metal grid filled in with glass panes. If you are on the top floor, or if the building is only one story high, skylights add to the space's luminosity. It is the quality of that light—not the views—that matters most. Quite often, windows will be translucent rather than perfectly clear, as the views in former industrial districts are sometimes not too exciting. Having an environment bathed in light is more important than seeing the world beyond the space.

The loft as a square and open unit of space is also as simple as possible. Nearly all the original loft conversions, in fact, created just one room in which the inhabitants figured out how to use the same space to sleep, work, cook, and socialize. Quite often, that meant treating furniture and furnishings as what the French call them: *meubles*, or moveables. Found chairs or beds could be placed anywhere, and curtains strung to create privacy. In that sense, the loft harkens back to the Middle Ages, when courts moved around the territories they controlled, bringing their furniture with them and installing the pieces in the large open spaces inside their castles. Nowadays, most lofts have walls that partition off certain spaces—such as bathrooms or bedrooms—to provide some privacy and even decorum, but even then, the divisions are minimal and sometimes do not go up all the way to the (inevitably high) ceiling.

The loft thus also represents the trend away from accumulating possessions and toward a leasing or borrowing economy (as companies such as Uber and Airbnb love to remind us). We surround ourselves only with things we need right now, borrow or rent what we want on occasion, and avoid amassing anything that might tie us down. As more and more media have become electronic or reduced to modular elements—so that what was once a room-sized computer now fits into the palm of your hand as an iPhone, or what was once a switchboard is now a router—that attitude makes sense. It also gives the few objects we have and how they were made more important. They even become the object of fetishes, as we have seen in the discussion of 500 Capp Street and its heirs.

Finally, the loft demands the presence of similarly flexible, large, and open spaces at the base of, or around the corner from, the converted warehouse in which it is located. Part of the idea of having a minimally appointed and open residence is that other such spaces are outfitted for socializing, shopping, or recreation. Some spaces are market halls, almost all of which have lofts above them—while others are gyms or bars. The same industrial aesthetic, with the implication of the conversion of work into play, defines the décor in such establishments.

Right now, the loft is the realm of those with a certain amount of money and flexibility, usually young people of means. Lofts in most cities are often expensive, though of course that differs from city to city. In general, some real estate sites define a loft as having at least a thousand square feet, which already makes them unaffordable to many. Moreover, as loft conversions generally are taking place in inner cities—once-declining housing markets, now coveted areas—they tend to be renovated, designed, and priced with an eye for a higher market.

However, the loft's flexibility and clarity could and should be used by a larger segment of the population—in particular by those with few possessions, changing circumstances, and a need for basic housing. They should, in other words, be social housing. I would argue that the obvious core stock of buildings for renovation into housing is our lofts, and that this form of repurposing has a strong tradition of imaginative reuse that we do not always recognize. I would also argue that we should learn from those traditions as we confront the largest task that looms in front of us: converting office buildings, often with large floorplates, into places to live. Making these spaces available for everybody is important—especially for those who lead transient ways of life or live with extended families not easily accommodated by single-family homes.

The possibilities of such conversions are, however, mainly still speculative. For now, the regulations governing social housing in almost every city in the world make it just about impossible to create the same kind of openness that marks the design of the best lofts. Social housing units must contain a certain number of bedrooms and other defined spaces, mainly because developers would otherwise try to cram people into places without the proper facilities. Conversely, the size of such units is usually also limited by the reality of economics: the government and the developers it hires want to invest as little as possible so they can stretch their funds to service the largest numbers of people.

Moreover, most social housing is highly standardized. It almost always consists of concrete plates or tubes (and sometimes wood in the United States) that are outfitted with mass-produced partitions, plumbing, and the other elements that make habitation possible at a bare minimum standard. These same practices mean that it is very difficult for existing buildings to be converted to new housing.

Until loft living is properly utilized across a greater variety of economic and social settings, then, it is to smaller residential structures that we must turn to see the best experiments in imaginative reuse for living. This includes apartment buildings and the most common form of housing in the United States—the single-family home.

HOUSING OUT OF RUINS

Many of the best of these experiments involve the inhabitation of ruins. To a certain extent, this is the result of current regulations. In most countries, you cannot alter what remains of a historic structure and, if you add something, you must clearly define what is new. If you restore the existing fabric, you must leave it largely as found, only reinforcing so that it does not crumble any further.

Moreover, the rough renovations of public and commercial buildings, as well as the rediscovery and reuse of industrial ruins, has made the idea of living in remains more attractive. The fact that the ruins are unfinished, have centuries' worth of character, and share a rich history with their surroundings makes them good anchors for new, loft-like spaces inserted within them.

There are examples all over the world of conversions of small buildings into apartments and single-family homes in which you can find yourself living in a ruin. A typically modest project, with a budget of about twenty thousand dollars, was the renovation of an abandoned house in the Spanish island of Mallorca by architect Carles Oliver in 2016 (see image #14). The renovation was sponsored by the local government to produce living units for inhabitants who service a huge tourist industry that has made the housing stock unaffordable. For that reason, and also because the house was a historic structure dating back to the eighteenth century, Oliver had to leave as much as possible intact, but he also delighted in what he called a "necessary archaeology."[1] Where he cut an arched door through a wall (as he goes on to say: "the most efficient way to make an opening"), he left the tile in place, removing pieces of it toward the top to reveal the stone underneath before letting it visually dissolve into a renovated plaster covering on the ceiling. That same succession of smooth stone innards, rough finished surfaces, and white walls continues through the spaces. In smaller areas where cuts were necessary, Oliver left evidence of where he had to cut into a wall to insert pipes or repair a floor. As in grander homes, the vistas through the existing structure open what was a collection of rather small and isolating rooms to the more flexible life of a modest apartment building.

In China in 2016, Zhang Ke redesigned and added structures to one of the "hutongs," or traditional courtyard districts, near Beijing's Forbidden City. The hutong was, and in a very few places still is, a rambling complex of brick courtyard houses interconnected with narrow alleyways to form a walled compound. Traditionally, each courtyard was inhabited by an extended family or clan, and would also include workshops and stores. Each individual room, however, was tiny, and the areas lacked basic utility services. Though this was once the main way in which most people in the inner core of Beijing and some other Chinese cities lived, the majority of such structures have by now been wiped out by redevelopment efforts and replaced with apartment buildings or other commercial uses. Only in the last ten years has Beijing begun protecting those that remain, providing subsidies for their renovations.

Ke, who had worked on several hutongs previously, used the courtyards as opportunities to perform their original function: to bridge the public and private and the domestic and commercial functions. He did so primarily by wedging plywood and steel structures into them to form new spaces. These new additions feature everything from a kindergarten to micro-residential rental units. The angles of the new rooms stand in contrast to the straight lines of the original buildings, but in these small additions Ke also mirrored the outlines of the old roofs with the plywood he inserted below them, extended brick walls into steps that act as stoops or play areas, and restored the original buildings to be housing—albeit not for the original residents whose families had lived there, often for generations, but for newer and wealthier occupants. From the outside of what remains of the historic hutong structures, you would not know anything had changed: all the layers of brick, corrugated metal, and wood that have accumulated onto the basic blocks are left in place. As you slide into the homes, smoother materials and more abstract forms begin to emerge, angled openings cut through the walls, and finally the completely new additions appear, rising up to clamp onto and over the tiled roofs.

GRANBY STREET

One of the most remarkable ongoing projects of imaginative restoration in housing is around Granby Street in Liverpool, England. The area, known as Granby Four Streets, is located in the densely packed neighborhood of Toxteth. In the second half of the nineteenth century and the first half of the twentieth, this area had been the domain of immigrants from the surrounding countryside and from Ireland, who came to work in the city's factories and shipyards. It became more racially mixed in the 1960s with the influx of the Windrush generation—the large migration of immigrants who came to England from the Caribbean, starting with the arrival of a ship with that name in 1948. Tensions festered that erupted into riots in 1981. Within a decade, at least half of the rowhouses in the area had been abandoned and boarded up.

In 2012, the London-based design collective Assemble started working with a local organization, seeking solutions for this situation. Instead of proposing either all new buildings or a more standard renovation of the stock that remained, they worked with residents to ascertain what qualities of Four Streets still worked and what was still needed. They acted not—at least at first—as designers but as activists and researchers: they looked for funding opportunities, fought for the rights of the inhabitants, and figured out what economic impetus could make the area come back alive.

In the renovations they did perform on existing houses, they opened rooms and extended them to the rear where possible. They always reused most of the materials they found on site, and left found objects as relics within the new walls they created. Assemble also realized they needed some additional building stock and used that as an opportunity for entrepreneurship. They set up a tile fabrication site nearby, Granby Workshop, to produce pieces for their bathroom and kitchen remodels. They trained local workers there and have gone on to utilize the products in other projects. Now the economic generators they have initiated are extending to other nearby sites—Assemble members have worked with local restaurateurs, market stall owners, and craftspeople to take over and refurbish appropriate sites.

The central showcase for many of these activities, as well as for Assemble's design approach, is another ruin. Called the Winter Garden, it appears from the outside to be just another rowhouse on Cairn Street, which runs at a right angle to Granby Street. Walk in the front door (see image #15), though, and you find yourself in a two-story space completely covered by a sloping glass skylight. In one section, the structure for what was once the second floor remains as an open grid within the space.

Inside this Winter Garden, plants luxuriate, including tropical vegetation that recent immigrants from Africa, the Caribbean, or Asia might remember. Their presentation also recalls the solaria and winter gardens that had been part of the neighborhood's grand villas during the same late nineteenth and early twentieth-century period in which the area's humbler tracts were built. Assemble left the brick walls in their bare and unfinished state, stripping away all the layers of paint and other coverings to show the rowhouse's basic building stock. They inserted steel beams, which they painted blue, to stabilize the remains. A staircase of similar color and material leads up to what remains of the second floor.

Offices and a kitchen are tucked into the rear of the house, and on the second level, in the area where Assemble did not remove the floor, there is a small apartment for visiting artists who are part of what is now the Granby Street Project—a community organization assisting the neighbors and organizing arts and education events. The new construction is sealed off from the open space that jams right into the brick walls, bisecting an old arched opening. In the rear, an old extension to the house now is an open area, covered only with wood beams and joists. Where Assemble did add new elements—in particular wood frames for doors and windows—they drew on their extensive network of craftspeople to highlight the new materials and how they are assembled, emphasizing the joinery and fabrication of those elements. The tiles—decorated in patterns that recall the floral imagery of the Arts and Crafts tradition but abstracted into swirls and whirls—add a similar touch to the bathrooms and kitchen.

ROTTERDAM ROWHOUSES

In contrast to the finished quality of the Granby Street Project, the renovations that Superuse Studio performed on a similar group of rowhouses in Rotterdam in 2019 is much rougher. They were faced with structures that had been not only partially abandoned but burned out in a fire. The demographic challenges, however, were similar to those in Liverpool: the neighborhood's population was a combination of people whose ancestors had moved into the city from the nearby countryside at the end of the nineteenth and the beginning of the twentieth centuries, when Rotterdam was an industrial and maritime powerhouse. More recent immigrants from North Africa and the Middle East have also moved there. In 2004 a group of artists broke into the houses and began squatting in them. Superuse Studio worked with them and local groups to renovate the rowhouses into forty-six new dwelling units, with workshops and stores on the ground floor.

Superuse's prime focus is, as we saw in chapter 1, on "circular building": constructing only with reused or repurposed materials, demolishing existing structures to make such reuse possible, and designing new structures so that they can be easily taken apart and their components reused elsewhere. At the Rotterdam rowhouses, which came to be known as W1555, they opened up what—through years of both legal and illegal renovations—had become rabbit warrens in order to create loft-like floors on each level. They then collaborated with the intended inhabitants—many of whom were the squatters who already lived there—and with others from the neighborhood and beyond to insert such new facilities as were needed, using recycled materials wherever possible.

As in the Granby Street example, very little of this new work is visible on the exterior, save for a few new storefronts and, if you look carefully, new windows. It is on the rear of the block that the transformation is most clear. There, Superuse ran new balconies and inserted stairs and elevators to turn the rowhouses into sets of what the Dutch call "arcade flats": residential units, typically one to a floor, stacked on top each other in which each is entered from an outdoor walkway running past what is usually the kitchen.

At W1555, the arcades consist of some new construction but also of recycled stairs, glass panes, and other structural elements that the architects culled from the rebuilding of the local subway system stations. They opened up the flats with large windows and doors to let light into the narrow spaces. Where they needed new elements—whether they were wood beams or roof covering—they used salvaged material. Even the concrete they had to pour was largely made by grinding up material from structures that were being demolished.

The basic construction was finished in 2019, and since then the inhabitants have added and subtracted elements, from fences to new additions, with Superuse studio sometimes being consulted and sometimes not. For Jan Jongert, the firm's founder and principal, that is "all part of how a building should evolve."[2]

The W1555 and Granby Street projects show a method of reuse that draws on buildings' original fabric, creates loft-like spaces, and could be extended into the realm of social housing. However, the approach is expensive. As Jongert points out, the cost of working in this manner was between 20 percent and 40 percent higher than if a developer had torn down what remained of the old buildings and erected new units in their place. It takes the commitment of the inhabitants and local activists, as well as a government committed to sustainable building practices, to realize that the added investment will pay for itself if you factor in the costs of environmental degradation caused by the extraction of raw materials for new construction. This economic conundrum makes such an approach almost impossible in countries such as the United States, which invests only the bare minimum into anything below market-rate housing. We would need to commit more in this area in order to adopt practices of imaginative reuse in housing.

There is also often a great deal of cultural resistance to this type of reuse. While wealthier clients might have seen examples of inhabited ruins in lifestyle magazines, fancy malls, expensive restaurants, or museums, those who have little access to such facilities may find this approach strange. Moreover, when you have felt hemmed in by the small social housing units, and by a past that is biased against you because you are in a less advantaged economic class or of a different race than those who have dominated your society, you may not feel much nostalgia about existing buildings. They may stand for a situation that you want to leave. As architects and activists seek to apply the lessons of experimental reuse to the area of housing justice, they will also have to learn how to open up and activate history's remains in a manner that reveals their beauty and potential.

TAILLIEU HOUSE

The places where architects have traditionally experimented—and where now you can find some of these forms of reuse pushed to extremes, showing how much even advanced ruins can be transformed—is in single-family houses. I came across a beautiful example in a small village in Belgium: a former farm building on the grounds of a cloister. The Belgian architect Jo Taillieu had fixed up the main complex, and in 2017 he had added a hundred units for assisted-care living. The outbuilding remained a ruin and was nothing but half a roof, most of the walls, and some wood structure. A couple found the ruin advertised for sale in 2018 and felt they could make a home out it, especially if Taillieu helped them. "We saw the possibilities, even if the current state of the place was rather scary," recalls the wife in that couple.[3]

Taillieu was constrained by local landmark regulations, which considered the structure—no matter that there was not much left of it—part of the overall historic context. So he left what he found still intact, then repaired and strengthened walls that had become freestanding because so much of the building's remainder had disappeared. Part of what had been a roofless room stayed that way, forming an outdoor patio sheltered by the remains of an outside wall, which screens the inhabitants from the public park that the cloister's grounds have become.

The architect's big move was to rebuild and extend what was left of the red tile-covered roof away from that outdoor patio to define the area below in which the new house plays out. That home is composed of both original elements and new ones. Below it, an added wall of floor-to-ceiling glass stretches up two stories, while fragments of concrete beams and walls help support the new structure. Taillieu emphasized what he had done at the larger scale of structural shifts and additions by painting some of the steel posts and beams blue and placing a galvanized metal strut to prop up one of the brick walls.

Inside, you can move in and out of what had been separate spaces through open doorways in walls, everywhere finding vistas cutting through the remains of the original building. New concrete floors, wood and steel beams, and a steel spiral staircase wind their way through the fragments of what Taillieu found on site. "Every time I sit down to drink a cup of coffee and look up," enthuses the husband of the couple who commissioned the design, perhaps unwittingly picking up on how generations of artists and writers have continually made discoveries in the remains of the past, "I see a new composition, a new way things are coming together. You never get tired here."[4]

That is not only because of the openness of the house, whose living areas ramble through what were once enclosing or framing elements with little effort, but also because of the way in which Taillieu delights in connections and contrasts. Each bolt that he used to attach a new piece of wood to a wall is visible, each crossing of new and old beams is celebrated by highlighting color and texture, and each slit and slot is framed to entice you to look through its aperture. The craftsmanship of these connections and intersections is highly refined, making for an even stronger contrast with the existing ruins.

Upstairs, where there was no remaining old material, the architect inserted wood cocoons that completely enclose the bedrooms, bathrooms, and dressing rooms that make up the more private parts of the house. The house luxuriates in both old and new materials: the working parts of the former building allow for a new life of upper-class leisure tucked away in a verdant suburb of Brussels.

ASTLEY HOUSE, PARCHMENT HOUSE, AND THE ANTIVILLA

Taillieu is not the only architect who has carved new, high-end living out of the remnants of old farm buildings, sheds, or homes. In England, in particular, a number of recent renovations celebrate the beauty and lessons of decay. One of most spectacular of these is the insertion of a relatively modest house inside the ruins of Astley Castle in Warwickshire, a grand manor dating back to the fifteenth century. Completed in 2013 to designs by Witherford Watson Mann, the project's priority was to stabilize and maintain what was left of the structure, which was mainly exterior walls. Inside a shell that once was home to a grand family related to British royalty (including Lady Jane Grey, the nine-day queen), the architects inserted a wood structure that contains bedrooms and living rooms. In those spaces, the wood beams attach to stone and brick walls, where the architects were careful to emphasize the way the smooth new material meets the rough old textures.

Even more spectacular are the spaces they left between the enclosed rooms and the existing walls. An outdoor dining room in what was once Astley Castle's great hall surrounds a wood table with a combination of old walls, including fragments of arches, and wood-framed windows. A skylight suspended in a new ceiling replaces what was once an ornate wood affair and sweeps over the dramatic, two-story remains of the hearth. Everywhere the contrast between what the architects left and what they added is clear, with Witherford Watson Mann opting for smooth plaster and wood to slide through the expressive forms of the ruins.

Slightly less grand, but even more expressive, is the Parchment Works in Northamptonshire, a similar effort at inserting a new home inside the remains of an old structure. In 2019, architect Will Gamble stabilized what remained of an old cattle shed and factory that produced the eponymous writing surfaces. Inside of those formerly crumbling walls, which still look like they are slowly receding to dust, he then placed a set of new spaces. There was more of the original wood structure remaining here, so Gamble cleaned up and reused the posts, beams, and joist, adding new supports where needed and painting the new walls white to emphasize the difference. He also had to add new steel structure in parts, as well as windows, and he clearly expressed these elements.

The Parchment Works also includes wholly new elements, including a rebuilt brick story in one part of the factory and a new building created out of bricks found on or near the site. The ruins are also more vestigial but, like Taillieu and Witherford Watson Mann, Gamble used the space between these remains and the new rooms to make outdoor spaces where you are sheltered from the outside but surrounded by the impressive evidence of the past.

In contrast to these rather romantic structures in the English countryside, the German firm B+ in 2010 was asked to convert a former lingerie mill southwest of Berlin, which had little charisma, into a single-family home. It was a concrete rectangle with an asbestos roof. Rather than trying to bring the whole building up to contemporary standards, the architects covered the hulk with shotcrete, a form of spray-on insulation. They used a concrete saw to cut holes in the walls to bring in light and offer views, leaving those holes in their irregular shapes. They then inserted a new core in the heart of the building, which contained all the sanitary and cooking facilities, as well as a new stair. They also left most of the remainder of the two floors open. Different areas are distinguished from each other with curtains, and the separate zones for eating, sleeping, and working can be individually climate controlled.

From the outside, the Antivilla, as B+ calls the result, is a gray box with some of its openings closed off with inset glass and some left open—in addition to the new ones they cut. A large downspout adds a note of drama but is necessary (although not at the scale the architects made it) because the old gambled roof was replaced with a flat one. Inside, the villa is austere, with all the surfaces remaining concrete and the curtains a white gauzy material. The Antivilla is as close to inhabiting a modern ruin as possible.

CHAPTER 7

AN ARCHITECTURE OF DOING NOTHING

GEHRY'S TEMPORARY CONTEMPORARY

"Don't tell anybody, but I didn't really do anything." That was the aside with which Frank Gehry drew me into his orbit the first time I visited his LA office in 1983. He was referring to the project he had just finished: converting an empty old car barn—where the city of Los Angeles had once stored trolley cars—into the temporary site of the Museum of Contemporary Art (MOCA), which had been founded a few years back did not yet have a permanent home. It was meant to serve as a stopgap solution while the Japanese architect Arata Isozaki worked on the museum's permanent home, which was to be tucked into the parking deck of a new cluster of high-rises. As it turned out, Isozaki's building, not only because of its design but also because of its location, would wind up being less than successful, both functionally and aesthetically, while Gehry's "temporary" building continues to serve the museum to this day.

Gehry's do-nothing approach was largely the result of his limited budget, but it was consistent with his architectural style at the time. Back in the 1970s and 1980s, Gehry delighted in designing buildings that seemed half-finished. He either exposed the wooden studs that are usually hidden behind drywall and paint, or else clad them with sheets of plywood or corrugated metal. He also extended his forms with chain-link fence, a material ubiquitous in the perennially sprawling, half-finished, and always redeveloping Los Angeles landscape. In his own house, which was a renovation and addition to a Queen Anne–style bungalow in Santa Monica, he had exposed much of the existing structure while leaving out the wall coverings to create a rambling, open living space from what had been a set of separate little rooms. My favorite touch there was the repurposing of the asphalt driveway into the kitchen floor—easy to clean, Gehry's wife Berta pointed out. Frank Gehry claimed at the time that Gordon Matta-Clark was his favorite architect.

So when Gehry was faced with the old car barn and a small budget, he was more than happy to leave the existing metal walls, brick and concrete block foundations, metal trusses, and metal roof not only in place but as exposed as he could make them, patching them up, replacing pieces only where necessary, and cleaning up and repairing existing skylights. He then considered what would be the absolute minimum needed to make the space usable as an art museum. New HVAC systems, plumbing, and electrical wiring used up most of the budget. Building codes required that he fireproof the columns and walls up to eight feet, so he clad them with the appropriate material and painted the new surfaces white. In addition, since the original car barn was actually composed of several buildings fused together, there were a few level changes, so Gehry created steps and ramps. Those additions resulted in a deck that became a natural place to gather and view the whole space. He painted the museum's name on the facade. MOCA needed a ticket booth, so he built a small shack outside, modeled on parking attendant booths, then made it more monumental by topping it with a back-lit chain-link box so it would glow at night. And that was about it.

The space was an immediate success when it opened in 1983, not the least because of the quality of the collections and exhibitions the institution had put together. But the building itself was also a draw. The white base of the new ramps and walls set off the rough textures, layers of wear, and mottled surfaces of the original building. The skylights revealed the intricacies of the structure above, and the viewing platform of the deck laid the whole space out before you. Not that this was necessarily something new: artists had for years been renovating their lofts with minimal means, doing only what was necessary and enjoying the beauty of the materials and spaces they found. Now, however, that private experience was a public one. And it turned out that the kinds of paintings and sculptures then in vogue—large canvases with strong graphic imagery, such as those by

Julian Schnabel, and sculptures with an equal presence, like the tilted steel pieces created by Richard Serra—looked great inside the open loft spaces.

The Temporary Contemporary, as the structure was then known (it is now the Geffen Contemporary, in honor of a major donor) was one of the first buildings to show how a space for art could also itself be a work of art by depicting what was already there. It showed off how it was made and what its materials and textures were. It did not pose itself as a neutral container, but rather invested (or infected?) the display space with its own history and nature. It was an anti-neutrality that critiqued the modern myth that an institution could be neutral and also the notion that architecture stood for something timeless and permanent.

This is not to say that the Temporary Contemporary was the first such structure. The first public art museum as we know it, the Louvre in Paris, appeared in 1793 by merely opening the doors of the royal palace to the general population so that everybody could admire the king's collection of art. Part of the attraction was that the paintings and sculptures were part of a larger scene of furniture, place settings, knickknacks, and heavily adorned walls that extended their artfulness into the objects of the royal household's everyday life.

The scheme was repeated in some of France's other palaces, such as Versailles, and was also applied in purpose-built showcases such as the National Gallery and British Museum in London, which opened a few decades later. Very quickly, however, the scientists and classifiers took over, developing what we now think of as museology. In that scheme, works were categorized into groups: paintings, sculptures, objects of use, and religious artifacts. How they were presented depended on their classification and on the scientific interpretation of each item. This meant that pieces were presented chronologically or according to type (still lifes versus history paintings, for instance), rather than according to how the original owner thought they looked best. The system was meant to be a machine for bringing people and art together, as were the labels, guides, and other modes of interpretation that evolved.

It was not until after the Second World War that the reuse of old buildings and the presentation of artworks became intertwined again in a way that—through their very design, rather than the arrangement and labeling of the works—questioned the role and value of each. When they did, however, they proved to be models for a kind of renovation that did not just put new uses into old buildings but also turned art into an integral part of the spatial and even structural experience and vice versa.

CASTELVECCHIO

The Venetian architect Carlo Scarpa's renovation of Verona's Castelvecchio, or Old Castle, is commonly cited as the first example of the type of renovation that displays, highlights, and elaborates the various parts and stages of historical buildings. In 1959 Scarpa was given the task of turning it into Verona's municipal museum. The original building was constructed between 1354 and 1376 and had been renovated almost continually, especially during the French occupation of the town in the late eighteenth century. Upon this foundation, Scarpa began a work of archaeology and projection that lasted until 1973.

If you visit the site now, it is difficult at first to see what Scarpa spent all that time doing. The main courtyard, which you reach by passing underneath a medieval arch, appears to be just a lawn surrounded by gravel walkways. Look carefully, however, and you notice that the stretches of grass and the pruned hedges are skewed and broken up. They trace and thus make visible, at least to the discerning eye, the outlines of former parts of the original castle torn down by the French when they occupied the site at the beginning of the nineteenth century. At their far edges, Scarpa created stone borders, beyond which the ground drops away to reveal actual physical foundations. A few concrete steps, framed in rusted steel, take you up just high enough to peer down into those uncovered ruins.

From there, look at the main facade facing you, and you will notice a few modern additions. A small concrete box peeks out, as well as a metal entrance door. To the right of that entry, Scarpa created a shallow pool fed by a fragment of a much older fountain. Look carefully at the front of the building, and you will notice bits where the smooth gray plaster gives way to a mottled texture. There, the architect has left some of the original wall materials, each a different color and from a different era so that you can understand the Castelvecchio's age and continuous occupation.

The most radical bits of architecture slide out from underneath the red tile roof at the building's far end (see image #16). The new roof, which mimics the original one in shape, shows itself as a restoration by giving way to the copper substructure built below it, until we see the wood beams and joist that support the whole. Below this awning, wood walls hang down in an abstraction of an older wall outline, while behind them a new wall closes off the space. In front of these excavations and restorations, Scarpa built a concrete stair hall, its many layers pushing back and forth, the stairs twisting around both new and old structures and itself, while a bridge connects two parts of the castle. Among this riot of the new and the old, the invented and the reimagined, the restored and the new, stands one of Verona's symbols: the medieval horseback statue of Cangrande I della Scala, the lord who built the Castelvecchio. Standing on a concrete pedestal that rises a floor and a half above the ground, he is cantilevered out for all to see from all sides. The history of the building, the town, and the construction all slip and slide by each other.

The inside of the building is filled with details that connect art and architecture, often quite directly. Scarpa was fond of making elaborate rails, hinges, and frames that turn walls, ceilings, floors, and structures into objects of use. The doors and screens with which he filled the old arches create a composition of rectangles that slide by each other, the walls, and windows, making for a second interior skin. That interior lining also includes new floors raised off the original stone and held back from the walls, as well as metal beams that double as structural supports and lighting tracks. When Scarpa breaks through the floor to show you the rough rubble below, he frames the opening with metal brackets from which wood rails dangle on chains. Everywhere you turn, the new is framing, focusing, and winding its way through the old.

When the architecture touches the art, Scarpa is equally obsessive about articulation. Some paintings stand in the middle of the galleries on black-painted metal easels, their backs there for you to see and admire as much as the Madonna that is depicted. Clamps, stepped and with visible bolts, reach around corners to support sculptures. A cross-shaped expanse of gray steel supports a crucifix. Swords and other armor are displayed in cases consisting of planes of wood and glass that barely intersect.

Throughout the Castelvecchio, you find the new and the old in an embrace. Sometimes it is a living and respectful one, but at other times one asserts itself over the other. A painted wall depicts the kind of textiles that might have once hung in front of it and is left as a fragment that destabilizes your sense of the wall. The gymnastics of the frames and clamps are more interesting than the rather mundane biblical scenes made by medieval and early Renaissance Veronese artists. It does not matter, however, whether it is the old or the new, or the art or the architecture, that catches your eye at any given time. It is the continuity of the craft that matters. The Castelvecchio is a display of making: the making of the original building and of the objects it holds but also of the welding of the building, new furnishings, and those pieces of art into a display. Things of different periods and made for different purposes are assembled to let you admire the ingenuity of the makers who built the historical and cultural legacy of Verona.

Scarpa applied this technique to quite a few other buildings, including a bank in the same town, a typewriter store in Venice, and several apartments. Most notably, he made renovations to the Uffizi in Florence and to two museums in Venice: the Ca' d'Oro and the Fondazione Querini Stampalia. The same tensions are present in all of them: between new and old, between structure and decoration, and between what you are meant to look at and what is meant to frame that object of attention. It was only in the Castelvecchio, though, that Scarpa was able to control just about every aspect of the work of art. Most museums could not afford to proceed with a twenty-year project. Scarpa's architecture also did not sit well with many art museum curators, who felt that his designs interfered with and even questioned the art objects they were trying to present. They were right.

SESC POMPEIA

A rougher and more easily accessible form of imaginative restoration was the one that Brazilian architect Lina Bo Bardi oversaw in São Paulo, Brazil. An early twentieth-century cooperage, or place to make wooden vats, became the leisure facility of SESC Pompeia between 1977 and 1986. Bo Bardi was asked by the progressive city government of the time to turn the site of the factory— a collection of low brick- and stone-clad buildings punctuated by smokestacks—into a complex of sports and entertainment spaces, workshops, a library, and even theaters. Instead of tearing the buildings down, as was common during that period, she decided to transform and add onto them. The physical and rather tall symbol of the project, akin to a church steeple, became the old chimney.

The project was not just one of renovation. Bo Bardi also added some significant new additions. In fact, what makes SESC Pompeia most visible are two concrete towers, the equivalent of twelve stories tall, one that houses stacks of gyms and the other changing rooms and showers.

Below these visual answers to the smokestacks and the housing blocks, Bo Bardi created a world of leisure by opening up the existing building. Most of the roof became skylights, supported by the original structure and new posts and beams. As much as possible, the architect kept the spaces open, using or adding walls only when necessary. She placed workshops in areas defined by both new low block walls and existing concrete ones, augmenting each with partitions from the same materials. Next to them she designed new wood reading booths for the library.

The most spectacular part of SESC Pompeia is where Bo Bardi left the space completely without divisions. Envisioned as a multiuse space and foyer for the enclosed theater, this space stretches underneath the light from above and over a concrete floor bisected by a meandering stream of water. That rivulet continues an existing drainage ditch on the outside of the factory in a way that helps to erase the distinction between inside and outside.

The sense of openness makes SESC Pompeia into the reverse of a factory space by subverting a structure designed to be functional for production into a place of play for what is now a community of readers, basketballers, performers, and people watchers who have flooded the facility since it first opened. The simplicity of the additions, which are abstractions of the elements Bo Bardi preserved, helps to create a new order that continues the old in the manner of a transformative archaeology or adaptation.

TATE MODERN

An even more radical transformation took place a few years later in London. In 1994 the Tate, one of England's oldest and largest museums, announced it would open a new hall of modern art in the shuttered Bankside Power Station. The power station had originally been built in 1947 to a design by Sir Giles Gilbert Scott and added onto in 1963; it had closed in 1981. The Tate initiated a competition for a design that would transform the power station's vast halls into a museum space. Herzog & de Meuron—a then relatively unknown firm based in Basel, Switzerland—won the competition, and the renovated building opened in 2000 as the Tate Modern.

The single most striking act of architecture in which Herzog & de Meuron engaged was the simple decision to leave the central turbine hall, a massive, brick-clad space whose large turbines had already been cleared out when the architects came on site free and open for exhibits. Not using a space that is over a hundred feet tall and five hundred feet long might seem like a perverse decision, but it turned out to be a stroke of genius. Visitors came by the millions as much to see that expanse as the collection and changing exhibitions. The Tate enhanced the turbine hall's draw by commissioning artists to create site-specific pieces there: one of the first of these, Olafur Eliasson's *The Weather Project* of 2003, drew over two million people. Visitors could just lie on the newly cleaned up floor of the hall to watch what appeared to be a glowing sun but was actually a trick of projection and mirrors in a space that was large enough to make this artifice appear convincing.

Within the part of the former power station not taken up by the main turbine hall, the architects added an escalator and balconies, but most of the other work consisted of cleaning up and abstracting what was left by emphasizing its simple geometries and planes: the power station's concrete frame was largely left exposed, as were metal trusses and gantries, while the floors were cleaned up and covered with new coatings. Within the galleries, new walls and technology took over.

Herzog & de Meuron did, like Bo Bardi at SESC Pompeia, add onto the power station. The most noticeable piece of new architecture is a glass rectangle that floats on top of the existing building where it faces the River Thames and, across it, the city of London. Announcing the advent of the new into this industrial site, its modest one-story scale appears more like a fissure, or accent point, on the behemoth main structure.

PALAIS DE TOKYO

In 2002, two years after the Tate Modern opened, an even more radical renovation made its debut across the Channel, in Paris. There, the 1937 Palais de Tokyo—originally built as part of that year's world's fair and named after the Japanese culture it displayed—reopened as a venue for contemporary art. Unlike the power station, the building was not heroic. It was a mess of meandering spaces because it was part of a formal composition of buildings—each clothed in vestiges of classical architecture—that had been designed to define a terrace across from the Eiffel Tower with little concern about how the resulting curves and angles might influence the interior spaces. For years, the city of Paris had used the building to house its collection of modern art, but in 1974 the newly completed Centre

Pompidou took over that function, leaving the Palais de Tokyo empty and neglected. In the 1990s, the local government decided to make it into a venue for contemporary art and experimentation.

The firm Lacaton & Vassal took an approach akin to Gehry's Temporary Contemporary: they did almost nothing—or so it appeared. They were faced with an almost completely empty building and, like Gehry, decided to add only the minimum needed. In this case, however, the building's varied spaces created a wider array of different scales and compositions than the open hangar that Gehry had faced in the 1980s. Moreover, by the 1990s the art that facilities needed to show consisted less of paintings on walls and sculptures on pedestals and more of performance, projection, and installation. This meant that Lacaton & Vassal had the freedom to do even less than Gehry had done.

Because the spaces were not so monumental, Lacaton & Vassal chose to highlight an aesthetic of drabness. They left in place not just the concrete structure but also the gnarly covering of fireproofing on columns and the racks of fluorescent lights stuck up against the ceiling. Only the walls and floors were cleaned up, with the patching and sealing left visible. There was, at least at first, not a white wall in sight—in fact, there were very few walls at all, and they were left as unpainted concrete or plaster. The only visible new elements were replacements of the original skylights and a few new structural pieces.

At times the Palais de Tokyo seemed to be a hairy beast, its layers of covering and fireproofing appearing to shed as you looked at them, parts of its air conditioning and electrical services dangling in midair, and artworks that looked to some like heaps of leftover junk piled up in corners. Rough signs, meant to show you where you were and where to go, came dangerously close to graffiti that acted as art. Furniture, much of it reclaimed, was strewn across the spaces, congealing to form a store or seating area, or dispersing itself across the halls for people to use as they saw fit.

Not all of the building was that messy. Most of the over eighty thousand square feet of exhibition area was open, high-ceilinged, and uncluttered, leaving it and its visitors space to come together. Windows and skylights bathed the rooms with light, even on cloudy days. The very meandering aspect of the design led you on through the various exhibitions, revealing new vistas and compositions of both the building and artworks as you curved around and up several levels. By now, different administrations and curators have cleaned and cleared up some of the spaces, and Lacaton & Vassal have added simpler renovations, but the clarity of the Palais de Tokyo and its roughness remain.

Part of the museum's success was the way it keyed into a changing aesthetic that had by then spread through most of Western culture. Born out of the punk movement, with its rebellion against authority and its love of the torn, the frayed, and the stolen (a book about the graphic side of the movement was called *Fucked Up and Photocopied*), it developed into a more sophisticated refusal to be beautiful in a conventional, canonical sense. A 1996 exhibition at the Centre Pompidou, *L'Informe: Mode d'Emploi*, became the touchstone for this new approach. The exhibition's curators, Yves-Alain Bois and Rosalind Krauss, used the very French, evocative, and therefore untranslatable title to suggest that they were interested in work that was not just informal but unformed or barely formed. Their show was meant to show art that was difficult to define as art because of its messiness, invisibility, or other conditions that made it hard to grasp or find. Including everything from Andy Warhol's drawings as a commercial shoe artist and Claes Oldenburg's relief of a fried egg to African cult objects made out of mud and leftover material, the exhibition highlighted art that did not want to be art but that aspired to be a condensation of, or perhaps just a found moment in, the chaos of the modern world. What it was, however, in all its minimalism, was concrete, or present, even in its inchoate form.

NORMCORE

This form of art making became embedded in—and some might say was a symptom of—a larger attitude toward everyday life that valued the normal over the new. By the 2010s, it had even acquired a name: "normcore." People, in a supposedly ironic manner, wore clothing designed for farmers, listened to music made up of song snippets that had been recorded decades earlier as hip-hop samples, and reappreciated the shopping malls, coffee shops, and schools of previous generations. Their goal was an aesthetic that was barely present but yet resolutely material and clear. It relied on borrowing and reusing with minimal changes made to the original. Over time, the techniques it used—such as straight copying of images and forms—evolved into editing images and later "kitbashing": taking found objects, scanning them, and combining them with new objects, textures, or colors such that you could not tell the different components apart. The results were compound images or forms, rather than shapes attempting to answer to functional needs, even if they could and would address those concerns.

This form of experimentation found a natural outlet at the fairs and exhibitions where architects were explicitly given the task of showing what they thought was the new or the future. At the most important of these, the Venice Biennale, they have begun showing not new pieces but the buildings themselves, most of which are located in a public park called the Giardini and are owned by the countries that organize their exhibitions there. For the last several years the German pavilion, for instance, has been under renovation, and so the artists and architects representing Germany have exhibited the excavation of the original building: they highlight the rough structure behind its Nazi-commissioned facades, and they show piles of construction materials as their display. Other architects who have been asked to exhibit in their country's pavilions have merely left their buildings empty so that you could admire such effects as the light sweeping across empty walls (the Belgian pavilion in 2012), or they have opened the pavilion to the surrounding community (Belgium again, in 2007, Austria in 2023). Yet others have created installations in which you seem to be in another place, like the recreating of a Syrian courtyard house by Mike Nelson in the English pavilion in 2011.

In 2010, the most radical and successful of these exhibitions took place, once again, in the Belgian pavilion. The firm Rotor designed an exhibition meant to highlight the importance of reuse, displaying fragments they had salvaged from buildings about to be torn down. These were not precious pieces of marble but linoleum tiles, scraps of carpet, and other fragments made of stuff that was difficult to identify but looked to be some form of petroleum derivative. The focus was on the things that you might try to ignore or rip out as soon as you got the chance to renovate it. Rotor hung the fragments on the wall as if they were precious works of art, complete with labels and careful lighting.

However, this approach turned out to be too radical to make it into mainstream architecture. Architects and their clients want the buildings that house art to be beautiful, and so far normcore has not, and might not ever, become the establishment's idea of beauty. Now the aesthetic of

L'Informe, like the more general idea of normcore, seems to have gone out of style altogether after only a short period, and we will have to await its rediscovery at some future date to see if it will then be treated as an aesthetic of a high enough quality (which would itself, of course, be ironic) to be allowed into the realm of cultural palaces.

NEUES MUSEUM

By contrast, the form of renovation that highlighted a building's heroic past and transformed it into a place of delight was almost immediately picked up by more formally minded architects who cleaned up the approach just enough to make it acceptable to a more mainstream audience—and to the clients commissioning the renovation of old buildings into culture palaces. In New York's Hudson Valley, the Dia Beacon—an old printing plant turned into a display space for modern art—appeared in 2003 as a cleaned-up version of the Temporary Contemporary. The Hamburger Bahnhof in Berlin, a former train station housing a private collection, did the same in 1996. Examples of this approach are now available in just about every city in the world.

It was the 1997–2009 renovation of the Neues Museum in Berlin, however, that really placed the rough-and-ready approach toward renovation in the canon of architecture. The monumental effect of British architect David Chipperfield's adaptation of this 1859 neoclassical building in the city's cultural core (the "museum island") has much to do with the restraints of the situation and the particular history of the place. The structure had originally been designed as a museum of antiquities and operated as such for almost a century. Its architect, Friedrich August Stüler, had attempted to evoke ancient Greco-Roman civilization. The frescoes in the main hall depicted a nineteenth-century version of some of the civilizations whose artifacts the museum then displayed. The building also bore the marks of the bombings and bullets of the Second World War. It had sat empty and in ruins until the renovation more than fifty years later.

When Chipperfield won the competition for its renovation in 1997, he first worked with a team of restoration architects to document every surface of the building. Computer analysis—as well as old-fashioned assessment with eye and hand—let him and his team decide which parts of the ruins they could leave as is, which needed to be stabilized, and which needed to be covered over or otherwise altered. The maps they produced as part of this effort were almost as beautiful as the resulting mottled walls that are part of the new building.

The renovated Neues Museum's walls, columns, floors, and ceilings form a visible history of the building that includes several layers of information and reference. They make up the bulk of Stüler's design that still stands. Some of the original frescoes are still in place, but the structures also bear the marks of the violence inflicted by both earlier renovations and, of course, the Second World War. Finally, the smooth stone that Chipperfield added at times turns into a modern evocation of neoclassicism, with new pilasters and cornice that are even more stripped down and abstracted than Stüler's forms.

For the elements that needed to be added to make the space a modern museum—stairs and other circulation devices, structural elements, roof covering, and mechanical systems—Chipperfield chose to abstract and shape them into rather aggressive modern forms. The central staircase became a cleaned-up version of the old one, while grids of columns defined new spaces in old courtyards. The contrast is striking, as this architect is not abashed in his scale and composition. The new elements vie with the old fragments of Stüler's architecture.

The clarity with which Chipperfield preserved the old and inserted the new made a form of reuse that had previously been acceptable only as a means to house contemporary art seem like a natural way to house art from all periods—like the antiquities that the Neues Museum displays. Chipperfield himself went on to apply the same technique, for instance, to the museum spaces that surround St. Mark's Square in Venice. It has become almost a trope now in art museums. Often that means not just making what was left visible in its often-fragmentary nature and contrasting it with new sculptural elements but also resurrecting the vivid paint colors and decorations that many such institutions favored before the twentieth century. A new form of museum has emerged: the art factory, evoking an industrial aesthetic whether or not it consists of a renovation of such a place of production.

POWERHOUSE ARTS

One of the most notable recent examples of an art factory is Powerhouse Arts (formerly the Gowanus Power Station) in Brooklyn, New York. The space was designed, starting in 2016, by Herzog & de Meuron—the same firm that created the Tate Modern. Their work at Powerhouse references one of New York City's most venerable contemporary art institutions, whose own design and history shows the development of this manner of artful reuse: PS1, which has since 1971 occupied a former schoolhouse of that name in Queens and which started as an independent contemporary art center. There was little money for major renovations, but also little interest in erasing the patina of use that clung to the classroom walls and corridors. PS1 basically took the school as they found it, cleaning it up where necessary, encouraging artists to make interventions and installations (some of which, like a "sky room" by James Turrell, became permanent), and using the meandering quality of the U-shaped building to create a labyrinth of art experiences. Eventually, the scrappy institution was taken over by Manhattan's Museum of Modern Art, which wisely left the building more or less as is, upgrading services only as necessary. Instead of expanding with a new building, PS1 created a series of outdoor rooms defined by unadorned concrete walls in the former courtyard, where it hosts a lively series of summer events.

Powerhouse Arts similarly seeks to encourage diverse arts activities. The site is Brooklyn's former Rapid Transit Power Station (which provided electricity to some of the public transport trains in the area), originally constructed as a steel frame building with brick panels in 1904 and decommissioned in the 1950s. Located in one of the most polluted industrial areas in the country, its empty forms, which were relatively easy to access, made it into an (illegal) haven for both exploring kids and tagging artists, earning the central space the moniker of "the batcave" (for an example, see image #17). In 2012, the billionaire Joshua Rechnitz, whom the press invariably describes as "reclusive" and who has funded several projects around New York City, bought the building and put more than $180 million into its renovation before donating the whole project to a nonprofit.

The most dramatic decision that Herzog & de Meuron made was to preserve almost all the graffiti that adorns the walls of much of the building. To a certain extent they had no choice: artists and their supporters threatened to derail the project if the pieces did not remain. The architects stabilized the walls, cut into them where necessary to create access, and painted the steel posts and beams a dark red that evokes the traditional colors of such components in factories and in the New York subway. They added wood floors, new ceilings, and a few other elements that contrast with the existing building.

Herzog & de Meuron also added a new building on the site of the plant's former boiler room. Here, the interiors are much simpler and dominated by concrete, with little of the historical layers of the power plant. The real action is in the main hall and other parts of this industrial relic, which has historically been not only a place to generate power but also a gathering site for New York's wayward youth and unsanctioned (and often untrained) artists.

There are thousands of such industrial relics across the world, most of which are destined to be torn down or renovated in a manner that preserves little of the original architecture, let alone the lives that took place in these buildings, the hard work and violence that made the turbines turn, the parties that were held there after the buildings stopped functioning, and all the other histories large and small that unfolded in these cavernous structures. That is because the type of institution that is best suited to display such a multilayered history is an art museum, and it turns out that most arts administrators still prefer their buildings to be bland boxes that supposedly provide neutral frames for the works on display.

Still, there have been enough renovations in the last two decades that the experiences at the Tate Modern or Powerhouse Arts is not that special anymore. The act of doing nothing to such rough and ready buildings—first demonstrated by Frank Gehry for a contemporary art museum more than fifty years ago—has become almost a cliché. It is, however, a productive one. Just as art often opens people's eyes to new ways of seeing their world, so the buildings that house that art are often the first places a general audience sees new approaches to architecture. The "do-nothing" approach has made socially acceptable the buildings that were long scorned as ugly remains of the industrial past.

CHAPTER 8

REUSING THE LANDSCAPE

THE HIGH LINE

In October 2018 I went to visit Manhattan's High Line, a mile-long elevated park that was developed over the last decade out of a closed-down freight line stretching from West Fourteenth Street to West Thirty-Fourth Street. It was one of those New York fall evenings when you realize that winter is coming. By 6:30 it was already becoming dark, and there was the proverbial nip in the air. This was also the magic hour, the time of optimal light for photographers: as the sunlight faded, the lights inside the surrounding buildings came on, outlining each structure in a glow from both inside and out, letting you see life extending from the streets through the bedrooms, living rooms, offices, and restaurants beyond the glass panes. It was the kind of time and place that could make you fall in the love with the city, and while at the same time you feel your heart break your heart as you sense the toil and the loneliness behind and in front of the facades.

I mounted the steps of the High Line until I was three floors above the street. As I stepped onto the deck laid over the old train tracks, I found myself surrounded on each side by singers, their faces glowing underneath caps with built-in lights. They were part of *The Mile-Long Opera*, an event put together by architect Elizabeth Diller and her firm, Diller Scofidio + Renfro, in collaboration with the composer David Lang and the poets Anne Carson and Claudia Rankine. "I'm walking toward what is mute. A neutrality of silences," the first group intoned. "Silence exists as an object that enables," they continued. The lines seemed a bit grand and abstract, but as I shuffled on, another story appeared: "She imagines she is walking toward a noisy dinner party, but, more often than not, it's dinner for one; it's just her sitting in her unbuttoned coat in the chair closest to the kitchen, eating the whitest plate and reading realty not reality." I instinctively turned away from them toward one of the apartment buildings surrounding the elevated walkway, expecting to see the subject of the song perusing the papers for better living quarters.[1]

As I moved farther down the High Line, the story kept changing and evolving. The table on which the woman was eating was described in more detail: "It was my grandmother's and I moved it from Elkhart to Minneapolis." The song traced the table's movement from the Midwest to Boston and then to New York. Along the way, the table was refinished and got new legs. Other tables appeared in the lyrics, some fancy and made of mahogany, others purchased from Ikea. The song touched on places from the Deep South to Slovenia. Other characters in the opera also moved in and out of focus, or so I thought; every time I stopped and listened carefully, there was really only one voice of one hybrid subject, all by itself, alone in their apartment and the city that was slowly settling in for the night around us.

Every time I stopped and looked at one of the singers, they would turn directly to address me, enhancing the sense that I was listening to one person's story. Then I would turn away and see the line of singers awaiting me ahead. I would catch shards of the story's next installment, and the lone singer and I would turn away from each other as I moved on. Beyond the singers, I would sometimes glimpse apartments that were more fully lit and see somebody standing there, staring out at space. In other apartments, window washers kept wiping the same pane over and over again. I realized the performance extended beyond the High Line.

Over the course of two hours, I walked the full mile of the park's linear space, then down the curving path around the towers of the new Hudson Yards development, a cluster of glass-sheathed towers then just rising up over the Penn Station railyards. As the space opened up, allowing you to look one way to the Hudson River and the other way toward the city skyline, the opera became louder, grander, and more abstract: "Whatever can happen to anyone can happen to us, whatever can happen to a city can happen to this city," the singers chanted. "The sleeping, the forgetting, the wrecking, the towering, the kissing, the scoffing, the cellophaning, the whirling snow, the sane and insane." They listed events small and large, from the applying of lipstick to "the rushing plenitudes," before ending with words that evoked two of New York's greatest writers, Walt Whitman and F. Scott Fitzgerald: "The silence after living . . . Onward rolls the broad bright current." As I moved past the last singer, I gazed at the Manhattan skyline one last time.[2] Then I descended a still makeshift staircase, made my way past the nearby bus terminal, and immersed myself again in the real city.

The Mile-Long Opera was a completely ephemeral event, though one that involved a great deal of logistics. It enlisted over a thousand singers from choirs all across New York to stand along the length of the High Line, and it organized the performances so that the piece began at 7 p.m. every evening of the opera's run—its subtitle was "a biography of 7 o'clock"—and descended into darkness two hours later. Along the way, the Mile-Long Opera sang life into the High Line and, with it, New York. It was the fullest embodiment of what that piece of landscape architecture had been able to achieve since its first section opened in 2009.

From a more brass-tacks perspective, the High Line (see image #18) has also been one of the most successful acts of value creation in the history of real estate. With an investment from donors and various governments totaling several hundred million dollars over the ten-year period it took to build all its sections, its construction unlocked close to fifteen billion dollars' worth of value: what had been an obscure part of the city was suddenly at the center of New York's cultural and commercial life. This formerly industrial part of Manhattan now attracted tenants to old buildings that quickly became co-ops and condos, and at least two dozen new towers were erected next to the High Line. The park draws about two million people a year, making it one of New York's most popular attractions. The western reach of Fourteenth Street at the park's southern end became a veritable high-end shopping mall, anchored by the Whitney Museum of American Art, which moved there from its old Upper East Side location at least in part because of the High Line.

This was, of course, not an altogether good thing from a social and economic perspective. Although the area had already been gentrifying before construction on the park started, it had still been home to many lower-income immigrants. One of the last diverse areas of Manhattan has since been washed away by a tide of multimillion-dollar apartments. As I walked the opera, I was briefly distracted by a group of what looked like frat bros having a party on their terrace, ostentatiously swinging around champagne bottles. What was once a working freight line had mainlined venture capital and tech dividends into the heart of New York.

The High Line, in other words, shows how reuse can unlock possibilities in both good and bad ways. On the one hand, it lets us experience the city in a whole new way and converts what had been a garbage-filled, abandoned track into an active public space. On the other hand, it has been the most efficient and effective tool of gentrification possible. These two outcomes, which are completely intertwined, are intrinsic to all the kinds of imaginative reuse that I have been describing, but their effects are particularly amplified in the case of reused transportation infrastructure. Around the world, converted railroad lines, canals, and even highways have become the prime movers of urban regeneration and reimagination. What were once industrial areas have turned into havens for tech companies, tech workers, hipsters, highly educated operators of symbolic logic[3] and centers for shopping and socializing—all by doing no more than opening and redesigning already existing spaces.

The High Line is the epitome of this. Previous attempts to provide new public spaces in Manhattan had yielded mixed results, with the changes to the West Side Highway that runs parallel to the High Line only a block away as a prime example. In 2001, after many proposals to tear down the still-busy elevated road, the section below Fifty-Ninth Street was finally removed. Many had hoped the site would become a giant linear park and a new way to bring waterfront access to Manhattanites. The road remains—albeit now a formidable, eight-lane barrier at ground level—and the few developments along the waterfront, though successful, have not transformed the area.

By comparison, the presence of the unused freight line in the middle of the city seemed to many like nothing more than a minor nuisance. After the last trains stopped running on it in the 1980s, less than fifty years after the line was erected, it was assumed that it eventually would be torn down. Then the usual pioneers of taggers, drug users, partying kids, hard sleepers, and a few photographers discovered the beauty of this overgrown linear oasis. They began arguing for its preservation. In 1999 the railroad company that had inherited the line, CSX, offered to donate the facility to whomever could take care of it. A nonprofit raised the money for the project, organized a competition for redevelopment proposals, and in 2006 selected Diller Scofidio + Renfro, together with the landscape architect James Corner, to design the new park.

The designers won the commission because of the balance they proposed between preserving what was there, recovering a sense of what had been there in the past, and creating moments that were new and unexpected. They worked with plant specialist Piet Oudolf to bring back some of the native vegetation that might have once flourished on the site but also to encourage some of the blow-ins (vegetation that had come to the site from the local neighborhood). New wood and metal walkways now thread between the tracks, making it easier to navigate the space, and turn up into vertical and then horizontal planes to become benches along the way. The original structure is still there, with new railings, supports, and stairs—all designed to be modern versions of the old structure and carried out in the same materials and modes of attachment. Diller Scofidio + Renfro and Corner kept these new interventions to a bare minimum—as is the case with the best imaginative forms of reuse—and were careful to preserve as many of the layers of use and patina of the old elevated line as they could.

What makes all this successful is the simple condition of the park being raised above the ground and located in the middle of the city. It sails over the streets below, arcing away from the Manhattan grid, leading your eye all the way to the Hudson River. Meanwhile, all around you are buildings that you view at mid-level, confronting you with their scenes from everyday life.

There is, as the opera articulated, a certain amount of voyeurism at play here, but beyond the lives going on behind those windows, you can also see what previous lives have built—there are hints in the bricks, panels, windows, cornices, and all the other details of construction around you. Finally, there is that sense of continuity. The space leads you ever on, distracting you at times with its views or the many performances it hosts or the people on benches, but always beckoning you into further spaces.

The moments where the High Line really sings, even when there is no opera or street performance going on, is when it literally slices through buildings. The cuts lead you into a cave where goods were once unloaded from the tracks inside the buildings they served, where graffiti remains, and where you can stop and rest. Then there is a moment where the wood planks veer off toward the east, step down, and form a small theater. Sit on the tiers and you find yourself watching the performance of cars whizzing by on Tenth Avenue below, in the never-ending automotive ballet of Manhattan.

OTHER ELEVATED PARKS

There are historic precedents to the elevated park—such as the Promenade Plantée in Paris, which opened in 1993—but the High Line's contemporary success has had a unique impact, inspiring countless other versions. In Chicago, you can find the Bloomingdale Line, for instance, which was first conceived in 2003 (about the same time as the High Line was beginning to form as an idea and organization) and constructed between 2013 and 2015. Nicknamed "the 606," after the beginning of the zip codes through which it runs, this park is also built on a disused railroad track, one that is almost three times as long as the High Line. The park is straight and has a simpler design than the New York prototype. The 606 also courses past lower buildings farther away from Chicago's core, so that it does not have the same sense of revealing the city and its lives. It does, however, connect to a much longer network of pedestrian and bicycle trails that snake up along the two branches of the Chicago River and its various tributaries and wetlands—the park directly under the final approach to O'Hare Airport is one of my favorites. It has also, like the High Line, caused gentrification: in the five years after it opened in 2013, house prices around it have tripled.

In Seoul, South Korea, the Dutch architecture firm MVRDV turned not a rail line but a former highway overpass into the Skygarden, which opened in 2017. It is a half-mile-long park with three fingers tracing what were once the road's on- and off-ramps. Recognizing that the history of the site was less romantic than the old freight lines—it was only built in the 1970s and consisted of bare concrete and asphalt—MVRDV restricted the greenery mainly to potted plants that dot the Skygarden's curves, although some of them are trees rooted in the earth. The designers hope they will grow to provide more shade. As the park is situated a few miles from Seoul's core, it offers only distant views of the city but has proven remarkably popular to flaneurs and children alike. There is something about being raised above the city in a space large enough to accommodate others that continues to attract.

The world's largest rails-to-park conversion, Atlanta's BeltLine, started in 2005, when the High Line was only an idea, albeit a well-publicized one. The BeltLine's construction is still ongoing in some sections. It has few raised areas but is as big as that city's sprawl demands: when completed, it will circle the inner core with over twenty-two miles of trails, plantings, and attractions. Its construction has proceeded in fits and starts, with many community members objecting to it because of the gentrification they fear it might cause and which has in fact occurred on some of the completed sections. The organizers and the city of Atlanta have countered by arguing that the BeltLine is part of a larger approach to the redevelopment of the neglected inner suburbs through which it loops: it subsidizes housing and transportation, hires local workers, improves public spaces, and commissions public art.

Compared to the High Line, the Atlanta linear (or continually circling) park reveals a completely different aspect of our urban condition: sprawl. Instead of encountering buildings in which you can see people living, working, and playing, you find yourself moving through the uncertain terrain between the human-made and the natural, and also among neighborhoods, industrial areas, empty lots, and patches of what is somewhere between forest and scrub. In the more open areas of the metropolis, housing is being produced in large clumps: four- to five-story wood-framed apartment buildings with between forty and sometimes well over a hundred units sitting on top of one level of concrete parking. Each of these buildings packs as many units as possible into dense masses clad with veneers meant to allude to either some historical era (Neo-Georgian fake-wood siding) or the modern age (steel and brightly colored stucco). This approach to housing has become the standard for cities across the country, and so in the Atlanta BeltLine, you can see how the American city is evolving.

DAYLIGHTING URBAN RIVERS

If one way of opening the city, of revealing both its past and its possible future, is to penetrate it at what you might think of as its waist level, the other is to dig into its base to find what it is built on. Cities around the world have been using this strategy to make both human and geological history present in the foundations of buildings, and to make the earth itself visible. In Seoul, for instance, part of the impetus for the Skygarden came from the immense success of the "daylighting," or opening to view, of the Cheonggyecheon River, starting in 2005. The water course, which runs right through the downtown area, had become at the beginning of the twentieth century a sewage drainage canal and was covered over with streets in 1968. A highway was built over it in the 1980s, and the river ceased to be used for sewage during that same period. In 2003 Seoul's mayor, after noting the need for public space, closed the highway above the river and authorized a new park.

Designed by the municipal engineering department, Cheonggyecheon Park is not particularly artful, consisting mainly of broad stone walkways meandering between walls clad in the same material. What makes it a pleasant stroll is the fact that the paving leads you down to the now-clear water in gentle steps, with no handrails or barriers between you and the river. The Cheonggyecheon—once one of the city's lifelines for transportation, water, fishing, and later sewage—is now a flowing presence, a place where you can dip your toes or fish, while the apartment buildings and office towers all around you rise up in closed ranks.

At almost seven miles long, the Cheonggyecheon Park is on a scale as massive as Seoul has become and has turned into a daily commuting route for many pedestrians and cyclists. Connecting the business district with several new cultural facilities, it shows off a metropolis that has grown in both economic and cultural weight. Walk along the park and you find yourself marveling at the scale of the place, but you also experience both its homogeneity—as its design reflects the rather repetitive form and materials of the buildings around it—and the liveliness of Korean popular culture, as concertgoers and art viewers spill out from venues and galleries.

The daylighting of streams and rivers in city centers has become a core tool in urban revitalization since the idea was first tested in the 1980s. Most cities were historically built around or near rivers, either to situate themselves at important crossings or to act as ports. As development and sanitation pressures grew in the nineteenth century, the rivers became foul effluents, and the impulse was to repress them by covering them over as much as possible. Their courses provided natural sites for highways, as well as flat, developable land. Only street names—like Manhattan's Canal Street—remind us of their former presence. And only the largest such water courses escaped the fate of being covered over—such as Paris's Seine and New York's East River. The massive infrastructure that was built alongside them—both to use their freight capacity and to protect people from floods—helped keep them out of sight and mind.

RIVERINE PARKS

One of the first efforts to bring a stream back into the life of a city was the recuperation of the Guadalupe River in San Jose, California, starting in 1989. This fifteen-mile river had helped irrigate what is now Silicon Valley. Historically, the Guadalupe had made the area attractive to both Native Americans and settlers, but its water came in torrents in the spring before turning into a trickle during dry months. By the end of the last century, the US Army Corps of Engineers had tamed the river, putting sections of it underground in culverts and building massive embankments around it. Meanwhile, the city of San Jose was growing because of the first tech wave of the 1990s, and the new workers and inhabitants saw the river only as a dark and dangerous void next to their buildings. In 1995 a coalition of conservationists and planners convinced the city and the corps to rehabilitate what became a three-mile stretch of the river running from downtown all the way to the local airport.

Designed by local landscape architect George Hargreaves—who has since made such efforts into a core part of his now international practice—the rehabilitation took place in phases and combined leaving what was there with adding new places. Hargreaves removed most, though not all, of the concrete retaining walls. He dug out parts of the channel to let the water flow deeper, then lined the bed with local rocks. He left in greenery that ranged from native live oaks to imported eucalyptus, adding new plantings to stabilize the soil and create a more continuous landscape experience. As the designers of the High Line later did, he also added amphitheater steps leading down to the water. In other places he created seating nooks, either by shaping the land or by inserting concrete and stone furniture. None of these additions appeared out of place, as Hargreaves was careful to follow the contours of the Guadalupe. Instead of making a new park, he coaxed a public space out of the natural course of the river.

The first stretch of Guadalupe River Park opened in 1999 and was an immense success. It encouraged urban development along its banks and brought people outdoors because of the shade and cooling that the riverbanks provided. Enterprising bicyclists commute to the airport along its line, and it turned out there were still salmon in the stream to be fished. What was especially remarkable about the park, at least when I first visited it in the early 2000s, was the mix of classes, races, and uses you could find there. Hispanic families held picnics next to office workers talking on their phones, while other folks practiced yoga or took a nap. It showed that San Jose was indeed a diverse city with a history rooted in its particular landscape.

By now, the park is beset with conflicts, as an explosion of homelessness and a lack of affordable housing in California has made it into a campsite, even while tech companies are planning massive redevelopments along its edges. After a quarter of a century of use and evolution, it needs a redesign, and a change also seems appropriate in a larger sense, given the manner in which water alters its surroundings as it courses through a landscape, whether that place is natural or human-made.

These days almost every self-respecting city around the world, and many smaller towns and villages as well, have daylighting projects either planned or in process. Some of them are small, like the planned rehabilitation of the small mountain stream in the town where I write this, while others are grander, such as Madrid Rio, the park that now runs for twenty-five miles through Madrid.

BIG FOOT LANDSCAPE ARCHITECTURE

The largest efforts to create such riverine parks are taking place in China, and almost all of them are designed by one firm, Turenscape. The firm's founder, Yu Kongjian, was born in 1963 in the small village of Dongyu in Zhejiang Province. Because he was from a "bourgeois" family, during the Cultural Revolution he had the job of tending the community's cows. He grew up watching how the animals moved, where they found water and shade, and how human beings had over the centuries shaped their habitats and fields to make the best use of water, drainage, and shade, and where the soil was most fertile and devoid of rocks. After attending Beijing University on a scholarship, Yu developed his ideas about what he calls "big foot landscape architecture": "The traditional Chinese garden is a complete invention in which nature is bound, in the way the feet of high-class women were, and can only survive with much support. The big feet of the peasants make landscapes that come out of how nature works."[4]

For the last few decades, Turenscape has been daylighting rivers and streams all around China. They have also landscaped the banks of major rivers to prevent floods, retaining soil where necessary and creating places where the cities' inhabitants can rediscover nature. The Quzhou Luming Park, for example, preserves and makes visible the sandstone hills that surround the park's shallow basin, while the main area is given over to working agriculture. Visitors can view the sunflowers and cornfields from elevated parkways. In a similar experiment, Yu turned the main quadrangle of Shenyang University into rice fields, giving students the responsibility of tilling the land as part of their education. Meanwhile, in the city of Zhongshan, Turenscape created a Shipyard Park that preserved relics of the old industries that had formed the site, including parts of hangars, warehouses, cranes, and railroad tracks. Where buildings had already been torn down, Turenscape outlined the foundations. They also created formal allées, or tree-lined paths, that cut through the natural setting to connect these fragments of the past. Between the paths, they designed new pavilions for shade and resting, and allowed lily ponds to recall the more classical traditions of Chinese landscape.

Turenscape is not as visually modest in their approach as many firms in the United States or Europe. When necessary, as in the Sanlihe Corridor, a two-hundred-acre linear park in Qian'anzhen, they will divert a river as part of an effort to clean polluted water naturally with weeds, which also creates a central element—a river along which you can walk—as an urban amenity. "Human beings have worked with nature for millennia," Yu points out. "The point is to work with it, not against it." Part of the goal is not only to learn from and reveal how nature works but also to understand what human beings have done on a site. Because a landscape surrounds you at a large scale, engaging every one of your senses, it can evoke not only layers of time and place that go back eons but also the human work and forms that have shaped it.

LIGHTER TOUCHES

A very different approach to landscape architecture is the extraordinarily light touch of the 2001 Strip Park, created by Italian architect Marco Navarra out a former railroad line stretching for twenty miles between several villages in central Sicily. Most of the work Navarra did there consists of no more than painting. With alternating stripes of red, yellow, and green, he transformed the asphalt path that had replaced the unused tracks into rhythmic elements that draw you through the landscape as you follow their pattern. Occasionally, hikers or bicyclists pass underneath metal arches.

Along the way, Navarra has made a few other interventions in the landscape. At a resting area near the top of a hill, he placed a metal step on top of a pile of rocks so you can ascend the boulders and look out over the site. At one of the former train stations, he made a small park by cleaning up the empty platform and adding a few walls and steps on which you can sit. He left the building itself as an empty ruin. The Strip lets you admire both the rough terrain of Sicily's center and the remains of the human efforts to connect the area to the coasts. It evokes a certain amount of nostalgia but also is a place that brings this empty area back alive.

Even more minimal is the intervention that landscape architect Jordi Romero made in 2011–2012 in the Turo de la Rovira neighborhood of Barcelona. The hilltop had been a site for agriculture before the Second World War, then for an antiaircraft gun emplacement during and after that war, and finally for houses built illegally starting in the 1960s. It also had a view overlooking all of the city. Romero, working with the architecture firm Jansana, de la Ville, de Paauw Arquitectes, preserved what they found. They removed only those parts of the illegally built structures needed to make the area accessible to the public, and then added only enough to allow visitors to use the area safely. Now the park is a continuous mosaic consisting of the floors of the various buildings that have existed on the hill, remnants of the military structures, the shell of an abandoned house, and new pedestrian pathways with concrete pavers winding their way through this mix. The contrast between the intricacy of the ruins and the sweep of the view is made all the stronger because of the absence of any elements to modulate between these two landscapes.

ILLUSTRATIVE LANDSCAPE ARCHITECTURE

Landscape architecture has a long history of being illustrative, representing Eden through "paradise gardens" (built equivalents of what the original paradise was thought to have been) and the humors of the body through various herbs. It also is a way in which human beings have realized a vision over a terrain that is larger than what they could cover with buildings. What is new is what I would call an imaginative landscape architecture: like the previous example, it works with what is, or was, there and tries to use the processes by which nature grows, changes, and evolves to make the relationship between layers of landscape visible even as it adds elements that open that ground to new possibilities.

Another version of this tactic is especially clear in Philadelphia's Franklin Court, on the site of Ben Franklin's long-ago demolished house. It is a form of urban landscape design through excavation and evocation that resembles a cross between an archaeological museum, a public plaza, and a large-scale sculpture. Designed by the firm of Venturi, Scott Brown and Associates and finished in 1976 (in time for the Bicentennial celebrations), the court's most visible part is the three-dimensional white steel-frame outline of Franklin's historic house. The effect is startling. Since Franklin Court is part of Independence National Historical Park—a section of the city preserved or recreated from the colonial era—all the details of the old buildings, including windows, doors, cornices, lintels, roofs, and stone bases, give these structures a scale. The Franklin House, on the other hand, is an abstract piece of sculpture as tall as the original building—close to four stories—and, without any details, appears both massive and light. The reality of where this Founding Father lived is there, but the structure has nothing to do with his actual life, just as our myth of him has little to do with his actual daily activities. Below the steel structure, the rooms are outlined on the ground in different paving materials, and you can peer down through glass scoops that act as viewing portals into the archaeological excavation that reveals what the actual house—or at least its basement—might have looked like.

These diverse examples—ranging from remolding and revealing the landscape, as with Hargreaves and Yu, to evoking that landscape through tracing and tracking, as with Romero and his collaborators, to reconstructing and marking it together as with Venturi and Scott Brown—make up the core of the imaginative reuse of landscape spaces. Landscape architecture and the deliberate making of public monuments have found many ways to mark existing spaces with new meanings, some of them more successful than others. It is the projects that most directly preserve what is there, in as many layers as possible, including those nature has made, that render the best results. Sites that merely list names, quotes, or years on wall labels usually do the opposite: they bury reality under the abstractions of language and art.

RE-INHABITING LANDSCAPES

Far more interesting, and at the very edge of what I would consider imaginative reuse, is an urban planning strategy that seeks to actually re-inhabit—rather than clear and redevelop—existing landscapes. This approach "fingers in" new development with existing structures, either within the city or, more usually, beyond its boundaries in the terrain of sprawl. This strategy is a way of preserving existing towns, villages, and agricultural areas that are larger than a small group of buildings but that lack the import of industrial complexes; it also allows large-scale new neighborhoods to be built in between what exists.

You can find one of the first examples of this approach, which dates back to the 1980s, at the edge of Rotterdam, Netherlands. The city had been expanding eastward into farmland since the Second World War, when most of its downtown had been bombed. By 1984 it was looking to take over another swatch of farmland called Prinsenland. This was a picturesque agricultural site traversed by country lanes and small streams. Whereas in previous expansion projects the city had razed whatever they'd found to erect rows of flats and attached homes, in this instance the head of the municipal planning office, Riek Bakker, decided to keep as much of the existing landscape as possible.

If you go to the Prinsenland neighborhood of Rotterdam today, you can travel down one of several main east-west routes. The straightest and most obvious one is the newly constructed Jacques Dutilhweg, a four-lane straight shot that runs by the main shopping district. Parallel to that road, however, are two old narrow roads that take longer to traverse: the Kralingseweg and the Ringvaartweg. Bakker's idea was to preserve not only these roads but also the structures along them, mainly farm buildings. Before the renovation, traveling along those older roads meant you would suddenly find yourself driving not on a straight suburban road but on a country lane, past open meadows interspersed with both centuries-old agricultural structures and new villas. These older roads were also built on dikes so that they gave you a sense of what it took to protect one of the lowest points in the Netherlands (up to twenty feet below sea level), while also giving you a vista of the meadows. You then confront the modern roads and apartment blocks again at the end of either of the streets.

By now, most of the farms have either been renovated or replaced by modern homes, and the meadows are also either built up or preserved as formal gardens. Still, the sense of the older landscape, which you can experience if you pick the roads up a few miles to the east, is present in the very layout of this newish neighborhood.

More than a decade after working on Prinsenland, Bakker applied the same principle to a much larger project—the establishment of a town called Leidsche Rijn near the central Dutch city of Utrecht. It was a twenty-year process that started in 1996. The new town was the largest of those planned by the government's VINEX program, which sought to find room for a million new homes to address housing shortages. Leidsche Rijn, with a projected population of close to two hundred thousand, was on track to meet much of that need, but it would also swallow up a big chunk of farmland, along with several historic forts, castles, and the old town of Vleuten.

Bakker and her team tried to preserve as much of the original landscape—whether natural or human-made—as possible, while still making room for all that new construction. Where possible, they kept the original rhythm of meadows running back from rural streets and, in one striking example, designed the area around Langerakbaan street so that one meadow would be preserved, the next one would be filled with two-story apartments, and the third would be left to agriculture. They also left several of the existing streets to run parallel to new ones— much in the manner of Prinsenland—and along them preserved not only farms but also sheds and industrial structures that included (according to one rumor) an illegal drug lab.

On a larger scale, the planners also preserved the outline of an agricultural area that once had been a Roman fort of which nothing visible remained. The planners' design allowed the local farmers to continue their work while adding a museum about the site and converting some sheds into a brewery, ironically sited right next to a brand-new mosque. The original "Old Rhine"—a small stream that at one point had been part of that mighty river's main course—was also left in place and surrounded by parkland. The Old Rhine goes on to flow through the much larger, new public Maxima Park, whose more than seven hundred acres are a collage of existing and preserved farms, farm roads, replanted forests, small built structures for displaying and making art, and paths connecting all of these disparate elements.

The remarkable success of Leidsche Rijn today is that, even though it is a medium-sized Dutch city with new blocks of housing, schools, sports centers, and shopping malls, visitors still can experience what had been there dating back centuries. Take one road and you are surrounded by the swooping forms of a school set against the repetitive geometries of flats; take another and you find yourself wandering down a country lane. Moreover, that latter road has not been preserved in amber. Everything there—from seventeenth-century brick farmhouses to additions built onto these structures in the 1950s to sheds put up last year—is a living landscape, showing off a continuity in time and place.

To my knowledge, no other countries have pursued this tactic of weaving new and old landscapes together, and even in the Netherlands the Leidsche Rijn has very few equals. It is, after all, a very expensive way to approach development that is in the short term more easily done by demolishing what is there to create the most efficient base for new structures.

ACUPUNCTURE URBANISM

In China, some architects have taken an analogous but somewhat different approach. It is what Xu Tiantian, partner in the Beijing architectural firm DnA, calls "acupuncture urbanism." Called in by the local municipality, the company renovates existing structures, adds a few new facilities, and then, with the towns' financial subsidies, encourages villages and small towns to restore what they already have.[5] In so doing, DnA works closely with the local authorities, neighborhood groups, and existing inhabitants. They used this tactic most effectively in the town of Caizhai, where they renovated an old agricultural building into a factory that produces tofu, a product for which the town is known. They then oversaw the rehabilitation of existing homes and the construction of new ones.

Similarly, the firm NEXT, also based in Beijing, oversaw the historic preservation of Dafang Village, one of more than a hundred abandoned or semi-desolate settlements in the area. Most of the population has moved to big cities to pursue the opportunities that they offer. The existing homes, stores, and workshops were carefully preserved—some of them in their decayed shape, some of them newly stabilized and fixed up, and most of them a combination of both. NEXT then inserted not only new facilities—such as museums, which they hope will draw visitors and eventually new inhabitants—but also new walkways, open areas, and a lookout tower from which to view the whole village.

One of the most successful such redevelopments in China is that undertaken by the firm Amateur Architecture Studio in the village of Wencun. This community hugs the banks of a small river in a landscape celebrated in classic Chinese scroll paintings and has managed to hold onto much of its existing structure. However, by the turn of this century it was emptying out as its inhabitants moved to China's booming big cities. When in 2012 the local government invited Amateur Architecture principal Wang Shu and his team to design a new attraction, which is how most towns seek to arrest their decline, the firm instead surveyed not only the village but the whole landscape around its boundaries. They traced the contours of the land, documented buildings both in use and ruins, and explored how the whole town could become a better version of itself.

The architects aimed to enhance what Wencun already was—a picturesque collection of structures that showed their construction over centuries—in order to attract new visitors and inhabitants. Their priority, therefore, was to leave what was there in place, accepting what had been built in all of its diversity while adding a few elements to continue the compositions and stabilize aspects of the buildings.

Amateur Architecture then inserted new buildings where they felt they were appropriate, using a combination of recycled tiles, stones, and stucco that they made locally. A few wood beams and trusses peek out of the new construction, which has a more abstract and formal appearance than the older buildings, but otherwise they fit into the fabric. Beyond the edge of Wencun's built-up area, the architects then designed a street of new buildings to house twenty-four households. The structures are built around courtyards, and the density of the housing blocks is offset by gradations of public places—from inner courtyards to larger ones, to restaurants and stores to public plazas—that invite people to gather in large or small groups.

The intervention has been a success; all the structures are now occupied, bed and breakfast facilities are opening up, and new businesses are moving into the village to serve a stream of visitors. Amateur Architecture is now looking to replicate this method in some of the other villages in the area, and the firm believes they can apply their principles in other parts of China as well.

My hope now is that the momentum of the preservation movements—which have been active for half a century and have by now also spread to most of the rest of the world—will be strong enough to resist the latest wave of development taking place in our exurban sites. Since in recent years remote work has become more widespread among the white-collar industries, small towns that are in desirable locations or are in themselves beautiful have become focal points for new residences. In many cases, it is exactly their historic character, as well as the presence of nonurban and agricultural landscapes, that makes them attractive.

We can hope that the approach tried in the Netherlands or China might continue elsewhere. Whether this form of imaginative reuse daylights creeks or gives you a new way to wander through the landscape, or whether it allows the patterns developed by nature and humans over centuries to become the framework for new construction—all these tactics should respect, reveal, open up, and continue what we have inherited. All around me, though, I see new tract homes arising around cul-de-sacs on what used to be farmland, and luxury residential compounds rising next to historic downtowns. I hope they can be stopped; there are better ways of making the new from the old.

CHAPTER 9

THE WORLD OF IMAGINATIVE REUSE

Clubs, Community Spaces, Markets, Shops

Piece by piece, building by building, imaginative renovation is becoming part of our lives. We read books in former car repair barns around tables made from stacks of discarded volumes. We go to work in old post offices where there are still pieces of sorting equipment interspersed among the desks. We study in colleges housed in what used to be shoe factories, where we can still smell the leather and oil.

STUDIO 54

This moment has been a long time coming. I first became aware that a built environment could collapse time and space, when a friend of mine, the author and cartoonist M. G. Lord, managed to convince the doorman to let a few of us into New York City's famous Studio 54 in 1980. After entering through the heavily guarded doors, I found myself in the lobby of what had first, in the 1930s, been Hammerstein's Theater, then in the 1960s the site where Ed Sullivan beamed his show to millions of Americans. The ornately decorated bar that still stood in the lobby was now staffed by young men clad only in gold lamé shorts, serving watered-down drinks and overpriced beer. I stopped to admire some of the details in the wall paneling whose curlicues I thought were evidence of vaguely Louis XIV-style decoration—by now layered with black paint and stains about which I did not want to ask too many questions—before being dragged through the theater doors into a vast space filled with gyrating dancers.

What was remarkable about the disco was how little new construction had been needed to make it the hottest attraction in the city. There was a small DJ booth and some new bars, but the main structure consisted of metal pipes and strings holding a mass of lights and amplifiers. The lights came in all sizes and shapes and produced a dizzying array of effects, some of them forming virtual walls that outlined a house shape in the air, others pressing down as a ceiling of pulsating neon flashes right above the dancers.

But beyond that, the Sullivan Theater—as everyone called it before Studio 54 moved in and again call it now—was much the same as how the owners of the disco had found it. That was especially obvious in the balcony, where the original seats remained, as a place not only to view the crowd and lights below but also to engage in the kinds of activities that made Studio 54 notorious. After being shocked by what I saw out of the corner of my eye, I realized that it was probably just a more overt version of what had been going on up there when this had still been a theater.

"This is our new Versailles," proclaimed my friend as she surveyed the space, "all the arts are coming together, all the aristocrats, all the drama, all the money." I had to agree, although I also concurred with another friend, who proclaimed with a mix of exhilaration and fear, somewhere in the early morning hours as the scene began to climax: "It must be the end of the world and we're celebrating."

I spent most of the night looking, not only because I was a little too fearful to do much else but also because my architecture-trained mind was fascinated by the way in which all those high-tech light projections were flashing off the stucco flowers and scrolls on the walls. Equally fascinating were the ceiling with its faded original paint and decoration, the faded velvet upholstery, and the holes and repairs that marked the walls. Studio 54 had converted a theater in which you sat silently, looking at a stage framed by grand but barely visible architecture, into a participatory performance in which everybody was an actor and a watcher, where the architecture faded in and out of view and importance depending on the mood of whomever was tuning the lights to the music.

Throughout the 1970s and 1980s, discos and bars pioneered imaginative reuse not just in New York—where Studio 54 competed with the Saint (another former theater in which hovered a geodesic dome) and the Tunnel (a disused subway tunnel)—but around the world. Few of them elsewhere were as grand as New York's, but many did make use of their historic sites. In Frankfurt, Germany, in 1990, you could find one club in a former air raid bunker right in the middle of downtown, but only if you knew where to look; another in a 1970s style bar, complete with vinyl seats and chrome details, at the Frankfurt airport. The birthplace of electronic dance music was in a converted theater in Detroit.

These spaces formalized the occupation of the cities' fringes by people on the fringe—a movement that had been going on for most of the twentieth century, if not longer. Those who took over abandoned spaces and began fixing them up were queer people looking for places to gather or cruise, taggers searching for sites, punks wanting to make loud music, and artists on a quest for free space to work and live. Interior decorators and fashion photographers picked up on the look of these spaces, especially once the style had gained acceptance through the likes of Studio 54 and the homes of successful artists. Soon you could find photo shoots with models draped around abandoned buildings; lawyers began living in industrial hangars. Stylish magazines and even movies spread the word and the look. I remember the hero of the 1981 film *Diva* driving his motorcycle right into the abandoned building he occupied, moving through what seemed to have once been an elegant structure of some kind, before taking a bath in a tub standing by itself on the parquet floor of a room of undefinable character and age. I want to be in that space, I remember thinking.

INDUSTRIAL STYLE

Soon enough, I could be in a chic industrial-style building. Restaurants and stores began taking a new approach to their renovations, foregoing dropped ceilings and white walls and instead highlighting their steel beams, the rivets that held them together, and the concrete floors. In 1986 architect Nigel Coates, a charter member of London's punk scene, renovated a Tokyo garage into a restaurant and nightclub by highlighting its industrial elements, but also by adding copies of classical sculptures, painted ceilings, and draperies that nodded to eighteenth-century France. You did not know what time or even era it was or where you were inside Café Bongo.

Over the years, this approach has become more acceptable and widespread. It is no longer reserved for nightlife and entertainment but is becoming part of our institutions and community spaces. It has also lost a great deal of the exuberance and sense of experimentation that made these early venues so exciting. Architects are now able to restore and renovate in a manner that not only preserves layers of time and use such that we can experience them simultaneously but does so in a manner that is relatively cost-efficient and can be applied in a wide variety of situations.

One of the very best examples of this approach I have seen builds directly on, or rather in, the industrial heritage that museums have been utilizing for quite a few decades. Located in Tilburg, a formerly industrial town in the southern part of the Netherlands, is a library and community center called the LocHal. Opened in 2019, it uses what had been—like the site of LA's Temporary Contemporary—a former municipal car and bus repair barn, located right next to the city's main train station. The architects, a collective of designers led by the firm Civic Architects, left most of the building's exterior intact. They only cleaned up some of the walls, and they replaced the top of the glass wall that faces the station and removed its bottom as well, both to make the project visible and to make the inside accessible.

It is the LocHal's inside that is most spectacular (see image #20). The front of the space rises to the equivalent of four floors, letting the steel structure—with much of its patina—strut its stuff all around you. Down the middle of this atrium, a set of wood steps tumbles from a mezzanine. Some people use these stairs to access the reading rooms and stacks above, others read on them, and occasionally a lecture or performance takes place in front of them. Beyond that, another large space, which functions as a more formal place for community events, is surrounded by most of the original floors that once housed the equipment and offices servicing the car barn. In fact, most of that machinery is still there, with all its dials and levers visible. Stacks take up some of the spaces, while reading rooms let you sit next to and under steel beams, lit by a version of the original translucent glass, looking across curving dynamos at the rest of the building.

The LocHal combines various functions, ranging from reading and borrowing books to hosting meetings, social gatherings, art exhibitions, and performances. Some government services are accessible there as well. Not only is the original building fully there, suffusing you with its textures and scale, but most of the furniture—like the tables made by stacking up discarded books and topping them with a piece of wood—is also made out of recycled materials. Its downtown location makes it an integral part of Tilburg's daily life.

An aspect of the LocHal that blurs time and place even more is the fact that most of its public spaces are not climate-controlled in the conventional sense. While some of the offices have normal heating and cooling, most of the places where you might read or gather have no such permanent services. Instead, mechanical systems woven through the original structure turn on in specific areas when people are present there. Curtains help differentiate some of the spaces and focus the heat when necessary. The result is that you have a sense of never being quite inside or outside but being in an in-between zone. Not only does that enhance the sense of suspension that is so vital to imaginative reuse, but it also saves the burning of a great deal of fossil fuels. "It certainly is not perfect," said a young friend of mine who has used the space frequently to study. "It takes a while for the system to kick in once you are there," he said, "and even then, you usually need a good sweater, but that also makes you feel like more like a squatter, which I like. You are not so much part of the system."[1]

More lyrical in its approach, befitting its function, is the conversion of an old Barcelona theater into the Sala Beckett, which opened in 2016. The original building—first an industrial space, then a social club—now contains a center for performances, a school, rehearsal spaces, and a restaurant, all designed by the firm Flores & Prats. Encouraged, to say the least, by the strict local and national historical preservation ordinances, the architects took an archaeological approach to their work. They found a building that was largely in ruins but filled with elegant details. Fragments of painted and tile-covered walls remained, as did doors and windows that had been designed in the Barcelona version of art nouveau, a style that was marked by whiplash curve, distended and elongated lines of mullions, and pale colors painted on the walls.

Flores & Prats polished and stabilized those existing surfaces and details and then finished the walls around these fragments. They restored or rebuilt the original door and window frames. They left the evidence of past repairs and places where the walls had been filled in and added new materials and connective devices as necessary. They painted parts of ceilings to indicate where there had once been another form of covering, leaving a white plane to appear beyond. They cut arched openings through other walls and ceilings to open up the space. They left some spaces, like an upstairs rehearsal room, largely as they found them, with stained walls and ceilings, adding only a lighting grid. In another space they painted most of the surfaces green, both recalling the original color of some of the interiors and creating a unified environment. A black box theater and a white-walled performance space were placed inside the original fabric.

The effect is, like LocHal, another form of spatial collage, but this one has a more elegiac atmosphere. The faded colors, fragmentary materials, and the broken-up nature of the rooms—some recalling their former uses, others cutting through that containment—give you the feeling that you have wandered into the kind of abandoned space where the character in *Diva* lived but now is condensed and enlivened by new uses. El Poblenou, the neighborhood in which the Sala Beckett is located is itself a gentrified version of its former working-class makeup of small workshops and apartment buildings, and what Flores & Prats have done, above all else, is to fix the memory of that place and combine it effectively with the contemporary culture that has taken it over.

Both the Sala Beckett and the LocHal are public spaces, and one of the reasons the architects could make something so radical there is because their cities worked with the architects to make their rather elaborate designs possible under current codes and with the available budget. There certainly are quite a few other examples of such cultural buildings that serve a modern community while evoking older ones. In the United States, those cultural buildings are, however, often part of larger public institutions. The renovation of a former Carnegie Library in Pittsburgh into the educational annex of the Museum Lab of the local Children's Museum by Koning Eizenberg Architecture—with its distressed walls and careful cuts and additions—is a great example of such a project. However, what is remarkable to me is how many purely commercial buildings are now following this mode of remaking, even if not always in as radical a manner.

HOTELS

The type of place that has the best chance at creating a completely immersive environment is the hotel. By their very nature, hotels are often older buildings that have been renovated every five to ten years, and the grandest of them often bear the marks of the many different interpretations of the original building. To walk into the Ritz in New York or the George V in Paris is to have a sense of history in which you can luxuriate, even if much of what you are seeing might be the product of a recent redo. In the last two decades of the twentieth century, the work overseen in various hotels by the French designer Philippe Starck—including the Royalton and Paramount in New York, the Mondrian in Los Angeles, and the Delano in Miami—made such renovations more knowing, allowing aspects of various periods to coexist with new elements.

In the last decade, however, a type of hotel has appeared that delights in contrasting the old and the new without making claims for continuity. In 2019 I stayed at the Hotel Warszawa in Poland. The building was formerly the headquarters of Warsaw's branch of the British Prudential Life Insurance Company and dates back to 1933. Once the tallest building in Warsaw at sixteen stories tall, the structure became a hotel after the Second World War. During that era, it was a Soviet-style, social realist rabbit warren. In 2018 local architecture firm GBD renovated it by stripping the building down to its concrete bones. Unfortunately, the social realist aspects were lost in the process, but as I lay in bed, I could trace those bones in the shape of concrete beams and posts rising through my room before they disappeared into the walls and ceilings. Sometimes that ceiling itself was cut away to reveal the original structure beyond the new surfaces and systems the architects had added.

Even more spectacular is the Warszawa's basement bar and restaurant area. Spilling down two underground levels, it finds its way through columns that are much more massive than the slim affairs in the rooms upstairs. To make space for stairs and circulation areas, GBD aggressively cut through whatever structure they could, leaving foreshortened stubs gesturing dramatically in midair. They also retained much of the original boiler equipment that once heated the building so that you can drink and dine among the machinery. The whole scene is lit by a new skylight cut into the courtyard above.

For an even rougher renovation you can turn to the Hotel Circulo Mexicano, right behind the main cathedral in Mexico City (see image #21). When the architecture firm Ambrosi Etchegaray began work on the nineteenth-century building in 2017 it was a ruin, though famous for having once been the home of the photographer Manuel Alvarez Bravo. They restored the outside to the way it might have looked when it was first built but took a very different approach to the inside. There, they left crumbling brick walls as they found them, only stabilizing them so they did not continue to disintegrate. Digging a bit further into what remained of the building, they also found fragments of stone and rubble walls and uncovered those. They left columns and lintels of finished stone that were still standing, some of them in the rescued innards of the ruins, some of them against smooth plaster where they had to fill in the fragments.

Through these shards of the building's original fabric (or fabrics, as different parts belonged to different periods), Ambrosi Etchegaray then threaded new stairs and hallways. They left the courtyard open but covered it with a skylight. New ceilings contrast with terra cotta vaults that remain in some rooms. The floor is a combination of stone and concrete, as well as both new and old terrazzo surfacing. The most startling interventions are the rooms themselves, which the designers kept minimal. Though most of them consist of new plaster walls, stone floors, and wood-framed windows and doors, with only a hint of original brick or tile showing up in some of them, they are so sparse as to seem like science fiction cells, albeit very comfortable ones.

At the Waterhouse Hotel in Shanghai, which opened in 2010, designers Neri & Hu took a similar approach as they renovated the old Japanese army headquarters building into a boutique hotel. They left the three-story concrete-and-stucco facade as they found it, adding and cleaning up only where necessary. Only the new windows in what were empty frames, along with a large door and awning made from Corten (pre-rusted) steel, give you a sense of what is inside. That is, until you step back and notice a fourth floor, also made from that rusty steel, sailing over the original building as if one of the ships that frequents this waterfront locale had wandered on top of the structure.

Inside, the space is even rougher, with the bare concrete shell of the building opened for the public spaces and reinforced with visible new steel. Windows, which were once parts of rooms on upper levels whose floors are now gone, dot the walls of the first three stories. Neri & Hu cleaned up the original floors to create walkways leading to nineteen guest rooms, their interiors as minimal, recessive, and simple as those at the Circulo Mexicano.

Although these architects have gone to extremes to preserve and showcase these original buildings, the approach has, in a somewhat watered-down version, become an almost mainstream phenomenon. The Ace Hotel group, which was founded in New York and has expanded all around the world, prefers to renovate buildings that have enough existing concrete and brick to give at least their public spaces, and sometimes the bedrooms, a recognizable character. There is a reason for this tactic, as the hotels then reflect the culture of the makers and tech employees who work in repurposed warehouses and who Ace sees as its core audience. The roughness of the materials and the informality of their outfitting, often with found or repurposed furniture and building materials, gives the brand a sense of being rooted in a history it has largely invented.

The Standard Hotel chain, which began in Los Angeles in 1999, prefers to highlight another period: mid-century modernism. Surfing the wave of renewed interest in that aesthetic, they opened their first hotel in a former motel along the Sunset Strip in West Hollywood, California. The brand's designer, Shawn Hausman, retained much of the original interiors and exterior, although the scalloped line of hotel rooms defined by stucco walls and glass sliding doors do not display many, if any, notable design traits. The reuse saved them a lot of money but also gave the place a sense of having been airlifted whole from the year 1960. Part of that was an illusion, with many of the curtains and carpets new versions of older patterns, but they then brought the period back even more strongly by outfitting the hotel with modern versions of furniture that a slightly more high-class hotel of the period might have featured.

The Standard management continued this approach when they opened their London location in 2019 (they have half a dozen hotels as of this writing), renovating a brutalist building that had served as offices for the Camden Council from 1973 until 2014. Hausman left many of the fixtures he found in the public spaces and retained aspects of the original building, such as the waffle concrete ceiling and the curved windows that perforate the massive facade. He then elaborated on what he found with furniture and materials, saying, "We tried to take it over in a friendly way . . . almost as if California rebels had taken over a government building and made it more free-spirited."[2]

The most visible restoration of a building from the glory days of expressive modernism is the transformation of the former TWA Terminal at New York City's JFK Airport into the lobby, bar, and restaurant of a new hotel called the TWA Hotel, established in 2019. Under the direction of veteran preservation firm Beyer Blinder Belle, the original Eero Saarinen–designed building was restored—based on original drawings and vintage photographs—to look much how it had appeared when it first opened in 1962. Walking into the TWA Hotel today gives you a clear sense of the sweeps of concrete that once whisked visitors from the drop-off outside, past check-in, up the stairs, and out to the lounges where the sloping windows and uptilted roof gave them a view of the airplanes and tarmac beyond. All of that, minus the check-in counters and information desks, are still there, and there are now also places to eat, drink, or do some people or plane watching. Only a few additions—such as the seating in what were once circulation areas and a few jarring fixtures—punch the time machine-like bubble that is the orgiastic shell of the TWA terminal-as-hotel.

RETAIL SPACES

Hotels have the advantage of naturally being sites of suspended time and place, while enveloping you with the sense that you have entered another world. Retail stores have less time and usually less space to do the same thing, but a similar effect can be achieved.

When I was staying at the Hotel Warszawa, I was taken around town by a local architect and found myself inside what had once been Warsaw's main electrical station. It has now been preserved as a shopping center, its steel structure filled in with brick. It has been cleaned up a bit too much, but all its pieces are still in place. From the outside, the Electricity Center appears as a pristine memorial to a mechanical age, with the control booth that protrudes from its front face painted a bright blue to mark the interior's new function. Added vent stacks and an elevator shaft also mark the presence of the new. On the inside, not just the control panels but also bits of leftover multicolored tile floor and wall sections join the cleaned-up structure to keep at least some of the Center's old life present.

Architects APA Wojciechowski left the center of what had been the machine hall open and cleaned up its skylights so the interior now glistens under the sun. In the control room, they left all the panels there and fully lit. That includes the special lever for the city's Communist Party headquarters, which had its own dedicated electricity line. Unlike on the outside of the structure, they left intact all the additions and marks of wear and tear. They then threaded a new steel structure through the existing beams and trusses so they could support three levels of stores tucked into the Center's sides.

This municipal building turned shopping center is one of many throughout Europe and elsewhere. The American counterpart is almost the inverse: old US suburban shopping malls are dying, prompting the emergence of "ruin porn" on sites such as deadmalls.com but also renovations of such spaces into schools and community colleges. While most of these renovations are banal, I hope that soon architects will find ways to preserve and reimagine a space that was central to many of our lives, and particularly our youth.

Within the American city and its first rings of growth, however, you can find the more familiar process of converting old places of production into modern sites of consumption. Old factories and warehouses are being turned into a new kind of mixed-use center whose semi-public space (as it is privately owned) consists of shopping streets, either within the buildings or between them, interspersed with as many restaurants and bars and surmounted by apartments or work spaces. Perhaps inspired by the Chinese art districts such as 798 Factory and old warehouses converted into living spaces, these complexes have become among the liveliest points of activity in older cities.

Atlanta, for instance, has the mixed-use complexes of Ponce City Market and the Westside Provisions District. Of these two, the latter is older, having started in the mid-2010s when parts of the old wholesale goods district were renovated. The area became formalized as a development in 2016. The redesign of the existing warehouse structures by the firm Smith Dalia was mainly an act of opening and connecting. The architects renovated the whole area into one experience to wander through: from parking garages built partially into existing buildings, down open paths and over bridges, and through new holes cut into the brick, block, and concrete facades. Inside the buildings you encounter courtyards filled with restaurants, bars, and cafés. The district even continues on the other side of active rail tracks, with the newly constructed bridge offering you a view of the freight trains as they make their way to and from Atlanta's rail yards—one of the country's busiest transportation hubs—before plunging you into another welter of stores and bars spilling out of former warehouses.

The Ponce City Market is much larger, refashioned out of what was once Sears's biggest distribution center on the East Coast. At two million square feet, the area has a lot of room, and most of the top floors of the eight- and nine-story former warehouse are leased as office space to companies such as Google, with other floors taken up by residential lofts.

Architects S9 did all they could to emphasize the scale and bravura of the old building, isolating its concrete frame for you to admire, cutting holes into floors to create vistas and a sense of height, but also just letting the sheer scale of the place take you over. The resulting market on the ground floors is almost overwhelmed: some of the smaller boutiques and cafés are dwarfed by the surrounding space, although larger establishments can use not only their size but also the intensity of their displays to play off against all that dead warehouse weight.

The Ponce City Market is the paradigm of such districts, though it is not the first, nor, I assume, the last. Since these markets in the bases of old warehouses have developed, they have become increasingly focused on food, leading developers to create a subset of such facilities: the market hall or food warehouse. Typical of this development is a new food court and marketplace in Brooklyn's DUMBO district, established in 2019. Using a brick warehouse with a wood timber structure that once housed the Empire Stores, the Time Out Market—named after its media conglomerate owner—brings together twenty-four restaurants into a series of tightly packed aisles that provide large expanses of open tables where diners can sit. Here, the structure is exposed wood, and the usual trusses and skylights give the space the sense of being a covered food market that you can find all over southern Europe.

POST OFFICES

Another offshoot of the renovated warehouse idea is the reuse of former post offices. These facilities have become available on a vast scale, mainly because the work of sorting and sending packages—which has completely taken over the sector from traditional letters—now takes place in purpose-built warehouses near major transportation hubs. This leaves the inner-city facilities, which often sport ornate public spaces, without much function. Over the last few decades, former post offices have been renovated into everything from Apple stores (I know of at least four such conversions) to train halls (in the case of the Moynihan in New York) to yet more shopping malls (found all over Europe). Most of these conversions left little of the original intact; the working spaces available for conversion were typically bare-bones in their design, while the public spaces in which the architecture was concentrated were usually under landmark restrictions and could not be altered—which is still mostly the case today.

An example of a post office that does manage to reveal itself in reuse is POST Houston, a 2017–2019 renovation of the city's former main sorting and distribution hub into a shopping mall. The main building was originally a three-story hangar fronted by a taller administration building. Concentrating on the main structure, the Dutch firm OMA gutted much of what was left, leaving the concrete frame in place and unfinished. They created three large atria dotting the two floorplates, then inserted elaborate stairs into the middle one of those, recalling the baroque grandeur of buildings such as nineteenth-century opera houses.

Constrained by the demands the developer imposed on the project, OMA still managed to delight in the exposed concrete, leaving the numbers and lettering that indicated postal zones. To top the whole project off, they transformed the roof into a 2.4-acre park that connects to a nearby nature trail. Future plans include transforming the parking lot into an outdoor market.

If such large shopping centers can impress us with the scale of the economic and physical changes transforming the operations of industry and consumption, it is in the smaller retail establishments that the more fine-grained ways of mixing past, present, and future are most obvious.

The first architects who, in the 1970s, started slicing into shopping malls and their constituent parts—at the same time that they were developing those shopping centers themselves—were the members of the New York firm Sculpture in the Environment (SITE), led by the artist James Wines. Their work was, at least at first, a kind of simulation. Working with a then fast-growing big-box retailer called Best Products—an early rival of chains like Target and Kmart—they designed shops that appeared to be ruins or buildings still under construction. After Best went bankrupt in 1998, Wines and his team took their weirdness to interior design, designing Manhattan stores for the designer Willi Smith, hanging clothing from chain-link fences or pipes they'd found during renovation. Cinder blocks piled up to form display areas, and rough concrete surfaces disappeared into shadows as spotlights lit only the clothing. The scenes resembled an abandoned cityscape. Such a haunting ambiance was at the core of SITE's sensibility: the drawings they did for their projects often evoked cities fading back into or coming out of either forests or ruins.

Beyond the SITE designs, the more direct forerunners of the rambling stores, apartments, and art installations cited above were collections of boutiques fashioned out of warehouses, small office buildings, and even courtyards of larger structures. They were themselves hybrids of malls and stores. They began appearing in the 1990s and included 10 Corso Como in Milan, Colette in Paris, and the Dover Street Market in London.

These early accumulations of displays in nooks, built-out mini boutiques, and mixtures of fashion, books, design objects, and other artifacts made it their point to ramble. They took the meandering classic shopping mall's ability to seduce you to purchase random items, shrank that wandering path, and then threaded the pathway through older buildings. In the process they left the walls, windows, floors, and ceilings as they found them, sometimes hanging clothes from handles or pipes, and using the roughness of plaster or wallpaper to set off the textures, colors, and compositions of the clothes.

Of these multi-boutique stores, the Dover Street Market was—and is—the most elaborate. Spread out over five floors of a cast-iron building on the eponymous street, the market mainly features the fashion developed by its founder, the Japanese designer Rei Kawakubo, and her collaborators, but it also extends out to show the work of designers in various media she admires. Here, the architectural interventions are strong, including (at various times—things change fast in fashion) a hut made from found lumber, a chain-link fence used to hang clothing, a "ghost building" made of wire, and various other structures constructed out of found or repurposed materials. Within the tight spaces—which press in on you both with the presence of the original structure and with all the material for sale—you feel completely submerged in a world where the sheer variety of design makes it uncertain what is new, what is old, or, most importantly, what is finished or unfinished.

Since Kawakubo and others pioneered the look of collaged and unfinished compositions, the style has spread throughout the world. It extends to the thriving world of pop-ups, in which the latest shoe drops or the most desirable and expensive fashion can show up in a store that was just vacated or containers that were plunked down on the street.

All these commercial spaces are selling you something. They are using the confusion of time and space that this kind of renovation produces to suspend their work outside of its context. This allows these retailers to claim a continuity with the styles of the past and with the craftsmanship of those historical fashions, but it also makes you sense that new designs can recombine and remix with what has come before to create something new. The mixture of low and high materials (or mundane and sophisticated, cheap and expensive) and styles, from the denim in blue jeans to the silk of ball gowns, also extends the confusion between what is utilitarian—that is, a building's structure and its basic wall, floor, and ceiling pieces—and how these elements present themselves with careful compositions and highly tailored forms. The store is the place where the suspension in time, place, production, and consumption start to seep, through the clothing on display, into our actual bodies and daily lives.

CHAPTER 10

THE MONUMENTS OF IMAGINATIVE REUSE

(Built from the Monuments of Industry)

American city planners have banished most factories and other sites of production as far away from city centers as possible, leaving their pollution and built-in class violence to other regions, countries, and even continents. In recent years, cities seem to have amassed a glut of office space (office use has dropped by an average of 20 to 30 percent in major US cities since 2020),[1] and we are beginning to see office buildings turned into housing at a fast pace in cities all over the world—from Terminal Tower, an art deco masterpiece in Cleveland, Ohio, to One Wall Street, a former bank building in Manhattan. The places of work that remain in cities have tried to shed their sense of rote mass production, focusing on "team spaces" and offering amenities housed in playful structures, like the kind of interiors Google favors. The transportation infrastructure that connects all these places is also shrinking, as new technologies obviate the need for large rail yards and as we have decided, in many places, to remove the highways through the middle of our cities. As we've seen, these former highways are becoming places for parks, while canals that once carried goods have long since become waterfronts along which to bicycle or enjoy a drink. Former rail yards from San Francisco (Mission Bay) to Queens (Sunnyside) are becoming completely new neighborhoods. All these changes in our environment are marks of the transformation of our economy to one focused on consumption rather than production.

This is not a new phenomenon: the replacement of production with consumption as the central driver of our economy was noted as early as late nineteenth century by Thorstein Veblen, while the term "consumerism" was coined in 1955 by John Bugas, a vice president at Ford Motor Company. It took a longer time, however, for these movements to evidence themselves in our physical environment. Over the last two to three decades, we have been seeing our daily built environment catch up with the realities of our economy. Now, cities are less and less sites where you go to work and more and more places where you gather or go to experience things. As for our houses, we are converting our kitchens into open spaces and our bathrooms into veritable spas, making our dwellings into open lofts, mimicking, ironically enough, the flexible scenes of industry.

MONUMENTS OF INDUSTRY

Historically, the cities of Europe usually coalesced around churches or castles, and in the United States around courthouses, city halls, or central squares outfitted with monuments. If our new society of consumers has any true monuments, they are on the sites of former industries: the complexes of steel mills, mines, and processing plants that were once the engines of the Industrial Revolution. Many of these locations are now turning into veritable cities that combine various leisure uses, cultural facilities, and educational and research sites with the remnants of these industrial behemoths. There are sites like this all around the world. The oldest one, though, is in the United States. In 1975, what had been the country's only coal gasification plant, located in Seattle, reopened as the Gas Works Park, designed by the landscape architect Richard Haag. In a manner that presaged and may have inspired later efforts, Haag left the plant essentially as he found it, turning it into the centerpiece of a more conventional array of lawns and planted areas on what had been the works' polluted nineteen-acre grounds. The Gas Works Park allows visitors to trace the intricacies of the various containers and connectors that once performed the distillation of coal.

The other remarkable aspect of the Gas Works is the way in which its red-colored structures have become the backdrop for concerts by the Seattle Symphony and others, staged in an amphitheater-shaped slope in the grass facing these structures As you watch performances, the remnants of industry stand as reminders of a different era.

The largest American conversion of an industrial complex is the repurposed Bethlehem Steel facilities in Pennsylvania, established in 1857. When I was an architecture student in the 1980s, we took a field trip to the then still operating plant, where we were suitably awed by the immense spaces and the molten steel churning away in giant vats. Buckets lifted the liquid steel out of the furnaces, carried it on gantries over to forming channels, and poured it into molds with an explosion of sparks. Now the 1,800-acre site is the home to SteelStacks, which houses a local community college, business incubator, and performance space. Unfortunately, many of the mill's buildings have been razed and are being replaced by office and classroom buildings, while others have been renovated in a rather bland manner that does little to keep the history of the place alive. However, five of the original blast furnaces remain as reminders of what was once one of the largest steel mills in the country. Elsewhere, from Buffalo to San Francisco, remnants of such industrial complexes remain, still frozen in time, awaiting their repurposing—or demolition.

THE POST-INDUSTRIAL PARKS OF RUHR VALLEY

You can find the largest collection of these kind of industrial conversions in the Ruhr Valley, a site that, starting in the 1850s, helped grow the German economy into the world's fourth largest by the beginning of the Second World War. The site extends not only along the hundred-mile Industrial Heritage Trail in the western part of the country but all along an arc of extractive industries that swings out from the Ruhr, through a slice of the Netherlands, across the heart of Belgium, and into northern France. It is here that you can find the monuments that are the calling cards of imaginative reuse.

By the end of the last century, the Ruhr—which had once produced almost half of Germany's industrial output and over 90 percent of its steel—had become a black hole of poverty and neglect marked by abandoned factories, mines, and railroad yards surrounded by soot-smeared towns with few prospects. The area was also heavily polluted. Though there were several attempts to address these issues, the real start of the regeneration effort was the International Building Exhibition Emscher Park, established in 1989. Sponsored by the government of North Rhine-Westphalia in western Germany, the park was developed by a pack of architects from around the world who concentrated, first, on cleaning up and making accessible the waterways of the valley's Ems River system and, second, on renovating, restoring, and finding new uses for the industrial sites along those waterways.

Over the next few decades, hundreds of miles of waterways were cleaned, pedestrian and bicycle pathways were added on each side (often on the sites of former rail lines), and vegetation thrived. Public artworks in the form of sculptures—and, in one notable example, a room where you enter to experience carefully manipulated light, created by the artist Maria Nordmann—popped up on the way, as did "adventure playgrounds" (a specific kind of play area where kids are free to roam and explore whatever is on the site, rather than having to play with equipment designed for specific games). These new functions used some of the leftover steel structures, coal yards, warehouse ruins, and other fragments of the industrial heritage. Each municipality (there are dozens in the heritage area, from large cities such as Essen and Dortmund to much smaller villages) took charge of its own remit, with only minimal coordination by the regional government of North Rhine-Westphalia. This created variety in the projects, with some interventions being minimal and others turning into veritable fairgrounds filled with new buildings.

Next, the regional government turned to the empty hulks of individual sites—the factories, smelters, coking and cooling towers, mineshafts, storage sheds, and all the gantries, bridges, and conveyor belts that connected them. Some became outdoor museums of their former selves, displaying the history of the industrial scene. Others, like the Jahrhunderthalle in Bochum, used their cavernous spaces to host expos and performances with a flexibility (due to their immensity) that traditional theaters could not offer.

It was the larger complexes, however, that showed the possibility of a more comprehensive reuse. The Zollverein Coal Mine Industrial Complex, for example, is a 250-acre development in the city of Essen that includes the original 1847 mine and the rationalist structures of the adjacent coking plant. After closing in 1993 and sitting empty for a decade, with plans to sell the equipment to China falling through, the city of Essen and the state government decided in 2001 to turn the Zollverein into a heritage site.

When I first visited the area a few years after that decision, it was still suspended between ruin and reuse. The buildings impressed me both with their size and the expressive nature of their elements, which ranged from abandoned equipment and the winding 170-foot tower of the mining shaft to the sloping bridges across which coal was once shuttled, around sixty feet above the ground. Between these huge hulks—each of them bigger than anything you might encounter in daily life—trees and other vegetation had grown, with veritable forests rising over slag heaps and weeds obscuring the roads. Enterprising artists and activists had taken over parts of the buildings and their surroundings, establishing studios, planting vegetable gardens, rehearsing with their bands, and making sculptures. In between all these other activities you could find a beer garden, and food trucks served us tourists. Some of this had been sanctioned by the local government, while some of it seemed to have just happened with little oversight (an example of what some of it looks like is image #22).

Since that first visit, the Zollverein has become more organized but no less impressive. Landscape architects have pruned and shaped the vegetation that had blown in or grown up there, and added native plants, while also laying down paths along the old trolley lines to make the whole area more accessible. An art depot bulges out of one of the warehouses with an angular obtrusion covered in concrete. The old machinery is still there to admire, but fences prevent you from climbing all over its expressive forms. The beer hall has become a somewhat fancier restaurant, and galleries have replaced the studios.

The Zollverein's largest intervention is the installation of the Ruhr Museum in the former coal washing plant. Designed by the Dutch firm OMA, the museum's exterior displays a modern answer to the old coal conveyor belts: a long escalator, its interior walls glowing with red paint, that takes you up the equivalent of twelve floors to the top level. There, OMA has built in a few new elements that continue the slick, glowing look of that long tongue, but most of all they have left the machinery to stand for itself, inviting you to wander among the spools and gears, levers and control panels, and pipes and grated walkways, where you can admire shapes formed not by artists but by engineers. That is part of this museum's point: the heavy machinery that made our modern world possible possesses its own beauty that we can now enjoy for its aesthetics.

At this museum, you can admire the marvels of human ingenuity and the sense of history present in the displays. The machines and gantries are not removed from their context—they look as if they could function at any moment. They are still grimy and evidence layers of paint. Some of their signs tell operators what to do, and others to stay away ("Dangerous to Life!" is a frequent one). This equipment is deeply rooted in its site and interconnected in a manner that makes it impossible to contemplate each part of it in isolation, or as finished artifacts. Instead, the whole embodies and represents decades of life and work.

The museum proceeds from immersing you at the top level of the eight-story building in this old technology to giving you a more standard history of mining and industry on the level below. Then, on the first level it turns into a more general historical and cultural museum of the area. The lowest level, before you are spit out through beautifully designed stairwells into the rest of the Zollverein, is a space for special displays, which when I visited was showing off the illustrious history of the area's soccer teams.

The Zollverein continues to grow as the government takes over more and more of the unused buildings, having already converted some of them into facilities for a small business school and for research and development laboratories. It has become one of Germany's most popular tourist attractions, bringing in close to twenty million visitors a year since 2014, although the effect on the surrounding neighborhoods is not clear. Though the city of Essen boasts of the influx of money that the renovations have brought—an average of sixty million euros into the economy annually—most of the surroundings still look as drab as when I first visited two decades ago.

Over the last two decades, similar sites have opened all long the arc of mines and industrial sites that runs from the Ruhr through the heart of Belgium to Northern France. Just across the border in Belgium, the old mine in the city of Genk turned into yet another combination of museum, performance space, art school, and park between 2005 and 2010. Very much in the mode of what OMA did at the Zollverein, the Brussels-based firm 51N4E (who also worked on the ASIAT site) turned the mine's main building into a combination of art, history, and mining museum. Here, you are meant to experience the mines more literally: a one-way, carefully signed and monitored path takes you down underground tunnels (which have, unfortunately, been sanitized for inspection, thus removing much of the grime and grit that was integral to the way such places worked and appeared), before letting you climb up to the top of the winding tower there. In contrast, the art museum is tucked away as much as possible into the machinery rooms and warehouse areas of the complex. The result makes for contrasts that are, depending on your stance, striking or jarring.

When I visited, the rather delicate chairs designed by Italian architect and artist Enzo Mari stood next to green-painted control gears and in spaces overwhelmed by giant gears. For the curators, that was the point: Mari, they said, had been heavily inspired by industry and had sought to distill its forms into his art and design.

In the French town of Saint-Étienne, a mine dating back to the 1850s was converted into a mining museum in 1994. The administration and curators made fewer additions than their German and Belgian counterparts, leaving most of the building as they found it, even down to the miner's clothes hanging from hooks. Here, time has not been recalled or suspended but merely stopped and the remaining artifacts exhibited.

The success of the sites in Europe have shown that people want to experience old industrial sites, and so efforts are underway to create similar attractions in almost every country in the Western world, as well as many in Asia. Authorities in China have picked up more forthrightly on the possibilities shown by the Ruhr Valley efforts. Although they too have razed many industrial sites, as the country has pivoted from large-scale production to more research-and-development-heavy and laboratory-based modes of production, Chinese authorities have also in recent years worked to save more of their industrial heritage.

THE SHOUGANG

When viewers tuned into the 2022 Beijing Winter Olympics, many were shocked. Instead of seeing chalets and tastefully curving structures emerging out of mountain settings, as had been the case at most former Games, they saw ski slopes swooshing past cooling towers and ice hockey rinks embedded in steel mills. "China really is our dystopian future," one poster commented on my Facebook; "they seem to be showing off their industrial might," said another. In reality, what we were seeing was probably the world's largest and fastest effort at reusing an old steel mill, the Shougang.

The Shougang Steel Mill started operation in 1919, and in 1953 Mao Tse-tung decreed that it should be the largest such mill in the world. He ordered it expanded to a vast scale throughout the 1960s and 1970s and succeeded in his goal. By the 1990s, however, it had become a white elephant: though still in operation, it was not nearly as efficient as more modern and compact plants, and it produced a huge amount of pollution. Moreover, the city of Beijing was expanding to the edge of the mill's once-exurban site. In the lead-up to the 2008 Beijing Summer Olympics, the Beijing government closed the plant as part of their effort to clean up the city and its air in time for the games. More than a decade later, the city decided to use the Winter Olympics as an opportunity to build facilities, from athlete housing to sports venues, as well as a park connecting them—all of which would form the backbone of a whole new urban district. This district has indeed developed around the repurposed structures in and around the steel mill.

The project was overseen by Bo Hongtao of CCTN, a semi-governmental engineering and architecture firm. As in most such efforts, the scheme for the over two-thousand-acre site involved a combination of remediation, restoration, leaving alone, and adding new buildings. What was different here was the addition of the sports facilities, some of whose scale and curving shapes were able to answer back to the coking towers, blast furnaces, and those particularly strange, corset-shaped cones that act as cooling towers for steel plants. In between the structures, CCTN planted trees along the sides of rail lines and overhead pipes, and they turned the water storage tanks into ornamental ponds. Some of the old containers were drained and turned into exhibition spaces, and as you descend into them by stairs you can see the water rising up above you through narrow slots.

The Shougang District, as it is now called, includes the obligatory museum that shows off the site's industrial history, but what stands out about this facility's repurposing is that the architects have not been bashful in their approach to the existing structures. They have cut holes through concrete and steel alike so that you can admire the equipment from unexpected angles. They even cut off some of the pipes and connectors, both to let you pass through and to give you a chance to admire the truncated shapes. You can enter silos and look out through porthole windows.

Other structures have been cut into with even more abandon. A Holiday Inn consists of a new and rather banal box of rooms around an atrium, but its bars, restaurants, and meeting spaces are in a preserved part of the mill, letting you sip drinks next to a boiler or hide behind some of the pipes to have a private conversation. As in the museum, the attitude here is aggressive: cuts through the structure create vistas past fragments of industrial history, and holes in the surface provide sometimes frightening views over, down, and across other pieces of the plant.

Much of what is now present at Shougang, post-Olympics, is banal. The new apartment buildings are as forgettable and as much a waste of resources as most of those in the rest of Beijing. With the neighborhood still not completely built out, the parks remain somewhat empty, as do some of the offices. The 2020 Olympic facilities that remain are, like those constructed in 2008, largely empty. The promise to use the base of the cooling tower as an event space has not been realized. But, just across the street from the main Olympics headquarters is a whole other ridge of unrenovated industrial structures rising and falling in heaps of tubes and containers over open steel networks. Fenced off and waiting for redevelopment, it holds a promise not unlike that of the Zollverein.

ART FACTORIES

Beijing has also pioneered another way to reuse industrial plants. While I was visiting the city in 2003, a local architect whisked me away from my comfortable hotel near the Forbidden City to what was then the industrial wastelands at the edge of the metropolis. Our taxi dropped us off at the gate of a complex of factory buildings, some of them still in full operation. Winding our way past trolleys and carts, we found the entrance to one of those structures, and suddenly we were walking past art galleries and studios and then sitting down in a serene tearoom–bookstore tucked away into one of the largely empty factory's corners. The spaces took up parts of rooms framed by plaster walls and concrete columns that held up north-facing skylights set in semi-circular cowls running the length of the factory. Maoist slogans from the days when this had been a bustling factory were still legible on some of those skylights (see image #23).

The site, which I mentioned in my earlier discussion on ghost architecture, was 798 Factory, originally built in the 1950s by China's onetime fellow communist ally, East Germany, to produce arms. It was owned and operated by the Chinese People's Liberation Army, but by the late 1990s production had move elsewhere and the factory sat largely empty. Artists discovered the space, staged informal events and shows, and then began moving in. Luckily, the army saw the potential of what was happening and regularized the arrangement, charging the new occupants rent and fixing up parts of the building. What started as a few spots of art in an otherwise empty building kept growing until, by the end of the 2010s, studios and galleries had taken over all of the useable parts of the building and spilled into the surrounding areas. Restaurants, stores, a hotel, offices, and loft apartments also appeared on the site, turning it into one of Beijing's most active commercial development areas.

The attraction was the low price—although that has changed due to 798 Factory's success and the expansion of the city of Beijing—but also the nature of the buildings themselves. Created by German architects according to a modular plan, the designs were ultimately based on the industrial factories that American architect Albert Kahn had created for the United States, the Soviet Union, and Europe before the Second World War. Kahn developed his pared-down, efficient layout for assembly-line production by working with Henry Ford, and by the 1930s was responsible for over 90 percent of factory designs in both the US and the USSR. Framed in either concrete or steel, his buildings emphasized daylight (brought in through saw-toothed skylights), the flexibility of large open spaces, and walls filled in with brick and glass.

The qualities that 798 Factory inherited from this design—the clarity of the structure, the surfeit of daylight, and the openness of the spaces—set it apart. It was an easy move for it to become a factory of art. Today, 798 is not just an Art Factory but a whole "Art Zone" that brings high profits to the army, which still owns the site. The district has moved far beyond its history in other ways. No artists work there anymore, and what galleries remain are generally very high-end. There are many boutiques, restaurants, and other services that make the Art Zone more like a shopping district. Those stores and cafés, moreover, retain little if any of the building's original character, and in some cases are completely new structures. In 2016 the architect Zhu Pei, working across the street from the original group of buildings, combined new concrete boxes with remnants of original factory structures to create the CUBE Art Museum. Some of the original brick walls are still visible, butting up against smooth concrete, and in the auditorium, skylights reveal the original trusses.

While 798 Art Zone seems destined to move completely beyond the combination of art and industry that once made it so attractive, it has spawned a host of similar complexes. One of the first of these was the Overseas Trading Center (OTC) in the city of Shenzhen. Shenzhen, now a city of twelve million, was only a village when Deng Xiaoping decreed in 1989 that it would be the first Special Economic Zone. Its OTC—constructed in the first wave of development into a television manufacturing plant and storage warehouses—was quickly eclipsed by subsequent facilities and was considered an ancient relic by the early 2000s.

In 2005 the site's owner, an enterprising manufacturer, had local architecture firm Urbanus clean up some of the spaces, hoping for the same sort of development that had occurred at 798 Factory. Urbanus worked hard to retain as much of the original character as possible, although the fabric with which they had to work was much less expressive than that in Beijing. When they found remnants of the old machinery and controls, they placed a metal mesh over them to let them be present without having people touch them (as they were sometimes still working). The architects' major contribution was a collection of new objects and landscape elements that wove the whole complex together. Raised wooden walkways ramp up off the sidewalk, rising as accessible pathways that also allow stores to shelter below those areas. They then broaden to make outdoor gatherings possible in the public spaces around them. An added staircase resembles a curved snake covered with green glass. New balconies and walkways on the outside of some of the taller buildings give the concrete blocks a sheen of translucent glass and metal. Small courtyards pop up between buildings, sheltered behind new block walls built around an old tree that provides shade.

The nearby city of Guangzhou also developed several art factories in the first two decades of this century. There, you will find the Xiaozhou Art Zone across three renovated warehouses; Redtory Art Park, once China's biggest can factory; and Taigucang Wharf, a collection of maritime storehouses that functions as a miniature version of 798 Art Zone. In Shanghai, the largest 798 imitator is the 50 Moganshan Road Art District, or M50, established in 2000. By now, almost every city in China has such an area, and there is even one in Moscow: the Winzavod Center for Contemporary Art. It takes up seven buildings that once served as breweries and winemaking facilities.

TRAIN STATIONS AND MILITARY BASES

There are fewer such large-scale sites in Western cities, mainly because American urban factories were razed long before this trend could blossom. One area where we do see relatively large-scale reused industrial complexes in American cities is among former train stations. In the second half of the nineteenth century, US and European metropolises saw the appearance of large, iron- and later steel-framed expanses like Penn Station in New York, Euston Station in London, and Gare St. Lazare in Paris. Their frames were covered with glass to house the belching beasts of locomotives and their train of iron cars.

Modern trains are clean, low-slung, and have new signaling and control systems that eliminate the need for marshalling yards and switching tracks. In most cases, therefore, the yards behind old urban train stations are being torn down and covered with new decks on which new buildings can be constructed. That has happened in New York (Penn Station's Hudson Yards) and is now ongoing in Philadelphia's Thirtieth Street Station. In both places, tall new glass skyscrapers are rising over the former train yards to house offices, apartments, hotels, shopping malls, and some cultural facilities. I wish the developers had made more use of the existing structures.

The other great opportunity for imaginatively reused monuments will be, I believe, decommissioned military facilities. In the United States alone, close to two hundred such bases, which are often vast, have been or are about to be closed. In most cases, these sites are cleaned up to remove hazardous materials and their structures torn down. The potential to do something more with what exists there is tremendous. Old military facilities in port areas, like the navy bases in Philadelphia and San Francisco, have already seen this kind of reuse. In many cities these military areas remained in use longer than the factories, and in some cases have only recently become available for reuse. In other cases, their redevelopment is still a future promise.

These ports were often spread out enough and had lost enough of their equipment to the wrecking ball or scrap dealers that they have seamlessly turned into neighborhoods that blend in with the surrounding city. This has happened with the port areas of San Francisco and Manhattan, as well as at Navy Pier in Chicago. Exceptions include the Brooklyn Navy Yard, the Brooklyn Army Terminal, and the Navy Yard in Philadelphia, all of which have been transformed.

Of these, the most impressive is the Brooklyn Army Terminal, designed by the architect Cass Gilbert and completed in 1918. Located in the Sunset Park neighborhood, its central building rises eight stories tall around an atrium through which trains used to load and unload goods (see image #24). Lit by Gilbert's skylight and shot through by walkways, it is dramatic enough that it has become a favorite set for sci-fi films (most notably 2019's *Joker*), even as some businesses have begun to occupy its lower floors. The whole site comprises almost a hundred acres; several of its smaller buildings are in full use, while music festivals and fairs utilize the imposing structures as a backdrop for their activities.

The Brooklyn Navy Yard is much larger—225 acres—and much closer to the center of the city, located on the East River just across from Manhattan. Established in 1801 and decommissioned in 1966, it has become an attractive site for businesses, schools, offices, and labs. The demand is high enough that several new buildings have been erected on the site. Very little remains of the original industrial equipment that once made this an active ship building and repair site: a few cranes stand by the side of the water, and inside some of the buildings, gantries and hooks still hang from exposed steel trusses. The sheer size and stretch of the yard, as well as its proximity to the East River and the remaining fabric of the warehouses, serve as reminders of the maritime life that made New York into one of the world's most important cities. At least one of the buildings has been restored with care, making room for small firms and workshops to gather around a skylight central hall that is a small version of the Tate's Turbine Hall.

The Philadelphia naval site, located along the Delaware River, is even older and larger, dating back to the American Revolution and spreading out over twelve hundred acres. The navy used the yard until 1990, after which they began consolidating their services around adjacent docks. The site now includes stores and restaurants that together employ more than fifteen thousand workers. It also houses some college and university classrooms and labs and a charter school, and many of the original structures have been fully restored. As in Brooklyn, there is little of the original industrial fabric left, and the old buildings that do remain have been cleaned up to such an extent that it is at times hard to tell the difference between them and the newly erected structures that are designed to look as if they are old. The stores and restaurants try to evoke the history, using fragments of machinery or ships as part of their decor.

THE FUTURE OF HERITAGE

This seems to be the fate of industrial heritage in this country. While we have reused railyards, ports, and military bases, most of the buildings that we have inherited from the Industrial Revolution seem to be too expensive to maintain, too large to use in what businesses deem to be an efficient manner, and too far away from centers of economic activity to be attractive. There is, however, a taste for industrial chic, promoted by companies such as Restoration Hardware and Shinola. Their wares are limited to tools, furnishings, and watches that are small and slicked up, however, and thus do not confront us with the reality of what industrial production was. Even artists—after a brief romance with industrial imagery in the 1990s, as seen in the paintings of Lawrence Gipe, Matt Mullican, and Ashley Bickerton—have lost interest. Only the specialized world of "steampunk"—an artistic movement that imagines an alternative timeline in which steam power remains the sole source of energy—continues to mix the industrial past with today's fashions.

Yet I would argue that the relics of the Industrial Revolution that are allowed to live on have made that vision real and available beyond the confines of art. The danger is that these complexes, as they develop, will become ever more taken over by all the things we think we need to be safe and comfortable, such as endless amounts of railings, dropped ceilings to hide air conditioning, and smooth surfaces, thus eliminating any imagery that could be seen as overwhelming or even frightening. That process has already taken place at 798 Art Zone and at some of the sites in the Ruhr Valley and has completely overwhelmed most of the American examples. We need these behemoths as monuments, and I hope enough of them will survive as such—not just in an experimental architect's imagination, but in reality.

An even larger question is what will happen with our currently operating industrial buildings once they become old enough to be considered heritage. Most production now takes place in hangar-like structures that express little of how they are made and leave little room around the equipment, which has itself become ever smaller and more contained within hermetic boxes. Factories and warehouses are difficult to tell apart. The latest monuments to how things work—the huge boxes containing the servers and computers that are the reality of our digital clouds—are even more closed-off and enigmatic, even as they sprawl for millions of square feet. It seems difficult to imagine how these structures will ever offer us a way to understand ourselves in space and time.

Structures such as the Zollverein, meanwhile, represent a moment in time and place. They are the industrial equivalent of medieval cathedrals, ancient temples, and pyramids. Just as those religious structures have in many cases lost their direct connection to how and why they were built (even if some of them are still in use as places of worship, albeit with much smaller congregations than they were designed to hold), industrial monuments are relics that impress us but that are alien to our daily lives. They have served as models for how we can renovate other structures. To find architecture that does allow us to find ourselves at home in our continually changing modern world, we might have to look elsewhere: beyond buildings.

PART 4

BEYOND BUILDINGS

CHAPTER 11

FAKING IT

Constructed Situations

THE MOVIE SET IN THE PRADA FOUNDATION

At first it seems like just another arts venue in an old industrial complex—in this case, a former distillery near the railroad yards in Milan, seventeen buildings in all. The new occupant is the Prada Foundation, the nonprofit arm of the clothing company whose headquarters are next door. The architect is Rem Koolhaas of the Dutch firm OMA. The old warehouses appear to have been cleared out and cleaned up, retaining some of their original patina and showing off their raw concrete. Two new buildings house the additional spaces the foundation needed, and they are slotted on top and between the warehouses.

Then something odd catches your eye. As the clouds part for a moment and the sun shines down on the complex, one of the buildings lights up. It is covered with twenty-four carat gold leaf. Why is not completely clear. You decide to catch your breath by having an aperitif in the bar within the complex. Walk in and you are confused again. Within the armature of steel posts and beams and under a ceiling of terra cotta tiles, the walls are covered in part with wallpaper depicting neoclassical buildings. You recognize it as having been fashionable in Milan in the 1960s. The lower parts of the walls are covered with dark-stained plywood. Jukeboxes and pinball machines mix with more modern booths. A waiter wearing a white shirt, bowtie, and apron takes your order. You seem to have landed in a space that is at once a café from the years of the Italian economic miracle after the Second World War, a fashionable restaurant from the 1930s, and a swank new bar. Where and when are you?

The café, it turns out, was designed not by Koolhaas but by Wes Anderson, the filmmaker who delights in drawing you into spaces and times that seem vaguely old and vaguely of a particular place. However, small objects scattered throughout the space—like modern coffee-making equipment and some of the more contemporary-looking plates and cutlery—hint that you are actually in a nostalgic version of the here and now. The café makes Anderson's on-screen vision into something you can inhabit.

Wander farther into the foundation's buildings and the odd tics add up. Walk into one of the new buildings, and the walls in the staircase are made of green metal grates, behind which you can see what looks to be an unfinished surface where plasterers and painters have started their work but then took a break and never returned. Where new and old buildings hook into each other, you have a hard time figuring out what is a new concrete column and what is an old wall. All around you is plastic sheathing, exposed struts, and more unfinished plasterwork. Inside the gold tower, where each floor is devoted to an installation by a different artist, some of the old windows are blocked off with an outdoor layer of expanded metal mesh.

New mirrored walls reflect the old buildings, making the doubling effect even more explicit. Koolhaas has lured you into a world in which the question of what was made and when is difficult to answer. Koolhaas has commented himself on what he was trying to do here: "The Fondazione is not a preservation project and not a new architecture. . . . Two conditions," he clarified, "that are usually kept separate here confront each other in a state of permanent interaction—offering an ensemble of fragments that will not congeal into a single image, or allow any part to dominate the others."[1]

But that assumes you can tell the two conditions apart. I think the real achievement here is that you can't. Koolhaas—like his client Miuccia Prada, whose work includes a continual recycling not only of vintage styles but of her own designs from a few decades ago—is more interested in the state of blur.

THE MISREMEMBERED TAHA BUILDING

Walking through the Prada Foundation, I was reminded of a peculiar building at 168 Upper Street in London designed by architect Amin Taha in 2017. It is a four-story combination of shops and apartments that constitutes, as Taha puts it, "a misremembered copy of a lost four-story building."[2] That lost building, dating back to the nineteenth century, had been a terra-cotta-covered, vaguely neoclassical part of a larger complex that stood on the site until it was bombed during the Second World War. In the early stages of his work, Taha went to the archives to ascertain more precisely what the building had looked like. He discovered that a still-standing building down the street was a mirror image of the one that had been on his site, and he created a digital map of that structure.

What he then built, though, was not a reconstruction. While at first glance you might see all the elements of a nineteenth-century structure there—complete with pilasters, cornices, bays, and bands of stone—you quickly notice that many of these elements do not align with each other. Columns sit on top of cornices, but also on top of single panes of glass. A door that cuts through another column is made not from wood, as the original had been, but from a mysterious material that turns out to be a lightweight cast concrete. Taha used a similar technique to outline windows but leave them as "false" ones, without the windows they should frame; actual openings dot the facade at what seem like random intervals but are in fact determined by the interior layouts of the apartments.

Taha sees himself as continuing history and bringing back its fragments, but in a way that he claims is much cheaper and more energy efficient than true renovation. Rather than elaborately preserving bits of buildings or moving spolia onto the site, he uses computers to direct 3D-printing milling machines that fabricate light and flexible versions of the original—or his interpretation of it—out of a kind of artificial stone. The whole building looks like a fake, and a rather clunky one at that, until you notice the skill with which he has put together his misremembered object. This is not an example of faking it until you make it, but of delighting in and pushing forward the simulation. Taha claims that building in this manner lets him "form a poetry of any number of visual outcomes and therefore meanings. . . . It is really up to all architects . . . to tell whatever story they wish."[3]

A new possibility has opened in this work of imaginative reuse. Instead of just assembling buildings out of actual leftovers, you can cobble together the past out of memories, imagery, and digital reconstructions. You can, in fact, make it by faking it.

PHOTOGRAPHED MINIATURES AND INDEXED BUILDINGS

The basement of the Prada Foundation takes this work of simulation even further. There you will find an installation by the German artist Thomas Demand. His work consists of photographs showing scenes both unfamiliar and strangely familiar: you immediately recognize an image of the Oval Office, but another photo, which seems to be of a standard apartment, seems unremarkable until you read the label and learn that you might have seen this flat in the news when a terrorist was arrested there. What makes the work even more uncanny is that Demand doesn't photograph those sites himself. He builds miniature models of them in his studio based on publicly available images and descriptions and then photographs them. At first, you might be fooled, but Demand is careful to let the ragged edges of the cardboard show and to let light seep through the seams between the little walls he has built. You are meant to see that this is a fake and at the same time to admire the referenced place present while also questioning the associations you have with the aura that has built up around its form and contents.

This work, in turn, reminded me of the Dutch artist Marjan Teeuwen, who creates art out of buildings that are in the process of being demolished—whether for sanctioned development in Europe or illegal settlements in Palestine. Her work consists of assembling the fragments into new structures in which the piles of stone, concrete, or wood are neatly stacked to form an index of the former building, thus making a new version of the old structure. She then takes photographs, beautifully composed and lit, and lets the demolition continue.

Teeuwen's work is about how we make art out of what we see, what we remember, and what meaning we project onto places. It argues that there is no one reality to be depicted, but rather we continually construct a place and time that we inhabit with the aid of our body and our mind, our eyes and other senses, and our imagination. The resulting photographs preserve virtually, rather than in reality.

On a more basic level, questioning our ability to distinguish between past, present, and times of a mythic, uncertain nature has been posed not just by thinkers but by those who make the images that surround us every day, including the dream factories in Hollywood and elsewhere. We have become used to the idea of alternate universes in the movies, books, and comics. After decades of debates about what history really means or is, fueled by philosophical doubts about the nature of truth and memory, as well as the continual unearthing of forgotten or suppressed aspects of our past, we understand the malleability of history, not just as an abstract idea but as an active process that informs our political and cultural discussions. Buildings that we thought represented the undying values of our form of government—the White House, the Capitol—turn out to have been designed to represent how white men of means thought we should be controlled. Some of those same white men now want us to mandate that we go back to designing our tax offices and courthouses in that same neoclassical style.

All of this leads me to the last category of imaginative reuse that I see as an alternative to the mindless churning out of ever more buildings: fake architecture. Of course, this category does not address our most urgent need, that of sustainability. However, the same computer-aided animation and special effects that let Hollywood create realities that are unrealities (or is it the other way around?) are also entering into architecture. Already, when a client commissions a building, the architect will create a digital version of it that is so accurate it is sometimes difficult to believe they have not already constructed it. Put virtual-reality goggles on—as some salespeople will have you do when they show you luxury condos that are still under construction—and the illusion is even more complete. How much longer will it be until we can live in a truly bare loft, with only images projected on every surface, complete with the textures we would like to feel and even the scents that go with that place and time?

DIGITAL SIMULATIONS

When I was just entering into the field of architecture in the 1970s, I saw a presentation by a scientist who claimed that one day soon you would be able to wake up and push a button to decide whether you wanted to be in Versailles or a penthouse on the ninety-ninth floor of a skyscraper. It has taken a bit longer than I thought, but his prediction, I believe, will come true soon.

A generation of artists and architects, most of them under forty, have become fascinated by LIDAR (Light Detection and Ranging) imagery, a technology that converts multiple panorama photos of buildings into information clouds that can then be transformed into a three-dimensional, animated version of the original site—and any number of other graphic representations. Their work can lead in one of two directions. On the one hand, there is the Madrid-based firm Factum Arte, which has made a name for itself by being able to reproduce not just works of art but whole spaces with absolute accuracy. You can now walk through a series of rooms that replicates the tomb of the Egyptian Pharaoh Thutmose III, for instance, and see newly created versions of the original paintings on what appear to be their original walls. You can see every crack, every pit, and even the dust that has accumulated on them. All of it is fake, but it looks real.

On the other hand, there are those who use LIDAR technology to conjure up imagined versions of real structures. The Senseable City Laboratory at the Massachusetts Institute of Technology has used the technique to map a favela, or shantytown, in Rio de Janeiro. Though their aim is to use the vast clouds of information they have gathered to inform social policy decisions that would improve these neighborhoods, their map depicts the favela as a kind of fairy tale world, made up of glowing surfaces and intricate forms that might in reality be sheets of corrugated metal or even trash by the side of the street.

In their project *Geography of Ghosts* (2023) the sociologist Wanda Spahl and the architect Dominic Schwab collaborated in Bern to create a hybrid of LIDAR scans, video, and imagery based on the stories immigrants had told them about the places they'd passed through in their journey. Featuring people who are themselves barely visible to most inhabitants of Berlin, this digital work of art documents, speculates, and builds a completely ephemeral space out of fragments of a reality inhabited by different groups in different ways and often at different times (see image #25).

The next frontier might emerge out of the advent of artificial intelligence (AI) programs such as DALL-E and Midjourney, which make it easy to produce images that seamlessly blend different times and places. Cesare Battelli, a Madrid-based architect, unearthed a suitably obscure sixteenth-century painter who created fantastical scenes, then told his computer to make a modern city in that mode and to blend in images of machinery with some of the scenes.

Although the programs such architects use so far tend to achieve the assembly by creating a literal blur of edges—softening the forms and bathing the scenes in soft browns, blues, and golds—they are learning fast that they can draw on sources other than *Game of Thrones* or *Lord of the Rings* for their architecture (which, because of the material on which it is trained and, I assume, the biases of both the programmers and those who use it, dominated the early versions of the software). A generation of architects who are now in school or just out of it use such AI programs to design new forms that draw on their reading and viewing, often caring little whether the result is actually built. I hope they will incorporate into their work some of the experiments in which architects and artists have engaged in making fake environments—such as the ones I have described above—so that they will lose their quality of escapist fantasy and reuse some of our daily reality.

SCENOGRAPHIES AND MALLEABLE REALITIES

Even though such imaginary blur spaces only exist on a screen, we are already living in much more prosaic fake places that have little of the uncertainty, evocative quality, or beauty of the works of art and architecture described above. Fake lofts abound, constructed inside brand-new buildings and outfitted with door handles, furniture, and curtains from Restoration Hardware. Another example is the headquarters of Meta in California's Santa Clara Valley, which has an interesting origin story. In 2013 and 2014 the company's founder, Mark Zuckerberg, was visiting several architects, looking for a designer for his new headquarters. When he walked into Frank Gehry's Los Angeles studio—a rambling collection of desks organized into project zones between exposed steel columns on concrete floors, all under the corrugated roof of a former warehouse—he reportedly stopped, looked around, and said: "Give me this. I need a few million square feet." And that is what Gehry and his firm did, recreating their own studio at a gigantic scale.

This form of faking is not just a question of making a few forms or spaces that look old. It is one of scenography. This is something we also already do, even if we might not realize it. We create complete environments—which often include the clothes we wear, the books we read, the shows we watch, the music we listen to, and even how we speak—to spin out a continually evolving reality around us. What is more, we have little need for architecture to do so. The buildings in which we construct our scenes need to be flexible enough to accept our projections and arrangements, and they need to keep us sheltered, but not much more than that. It is another reason why we really do not need to build anymore: we can recreate whatever reality we want in just about any structure we choose.

I do believe that there is a task for architecture in this production of a malleable reality. Those bones we are building within, after all, are not architecture. They are buildings. Architecture is not building or buildings; rather, it is how we design them, how we conceive them, how we represent them, and how they transform into a real scene. Architecture is the meta of building. It uses buildings and is usually most successful when it does so, but it can also exist in art, in literature, on a screen, or even in our imagination.

To make architecture, in other words, you don't have to make buildings. That is something I have tried to show in the examples above, but here it leads me to a final argument for not building: what we need is a practice that could also be a self-conscious and critical way of making fake buildings. If we can make anything, anywhere, anyway we want to, architecture can provide the meta-fake: a way to continually question where and when we are, in a manner that draws us back to an awareness of our world, our fellow human beings, and ourselves.

The quest for such work in practice brings me again back to the world of art. While David Ireland, Theaster Gates, and others have remade places for particular purposes and turned them into art, others have stepped back to explore the possibility of creating a true blur or a fake real space. They have imagined, portrayed, and in some cases built fake worlds. I believe that their work might be building blocks for a form of fake imaginative reuse.

LIFE-SCALE PHOTOGRAPHY

Some the most beautiful experiments in this area have been undertaken by photographers. During the last decades of the twentieth century, the German artists and teachers Bernd and Hilla Becher specialized in documenting structures that were functional but also enigmatic. The couple trained their lenses on mine shafts, water towers, and medieval buildings. The images they captured were as straightforward and direct as they could make them—shot on days when the sky looked like a uniform white-gray wall—and produced to make every detail of the structure visible. In one sense, they were documenting building types, most of which had outlived their usefulness, but in another sense they were turning the buildings into virtual sculptures that have in fact inspired several generations of architects and may have contributed to the restoration of some of the industrial sites they photographed.

Even more intriguing is the direction taken by some of their students, including Thomas Ruff, Thomas Struth, and Andreas Gursky. They, too, created large-scale photos of their chosen structures, taken in a seemingly neutral manner. The more you looked at the images, however, the odder they became, especially after they started using computers to digitally alter the results. While Ruff photographed streets very early in the morning so that the perfectly normal buildings appeared as abandoned ruins, Gursky concentrated on active sites. With the same eye for detail, he photographed hotel atria, canals, airport departure areas, and five-and-dime stores, flattening out what he saw both in form and color and lifting these everyday scenes out of context.

In more recent works, both Gursky and Ruff have begun to deform and blur their images, letting the computer turn a racetrack in the desert into fragments of black lanes cutting through sand dunes, or pixelating images found on the Internet so the images become deliberately uncertain in what they are representing.

Other photographers go beyond shooting what they find in reality and instead build sets for their photos that go far beyond Thomas Demand's models. The American photographer Gregory Crewdson spends up to a half a million dollars putting together what amounts to a complete movie set, all so that he can take one photograph. His subjects are the former mill towns of western Massachusetts, but he does not concentrate on the unused mill buildings or the other remains of the area's past industry. Instead, he focuses on everyday life and its scenes, both inside the houses and out in their neighborhoods, which he manipulates into stage sets. His earlier work from the beginning of this century mainly included domestic scenes in a home filled with the kind of furniture and possessions a family might have accumulated during that period. The actors in the rooms stood or sat with rather forlorn expressions, as if they had just had a fight.

It is in the outdoor scenes that Crewdson's photographs open up while still maintaining the sense that what we have built around us is oppressing us. Crewdson shoots most of his images from ladders or cranes to distance himself slightly from the scene. *Redemption Center*, an image from 2018–19, hovers over an empty parking lot, either at dusk or dawn (you can see it image #19). In the middle of the image is a single light pole and behind the light is an old store that at some point has become a Redemption Center, or so the faded letters on its side would have us believe. Its curtains are drawn, and thus it is difficult to make out what is happening there, but the building does have a new awning that contrasts with the worn appearance of the rest of the structure, and its parking lot, tucked off to the left, is better maintained than the one in the foreground. The two cars parked by the building are 1980s models, making you wonder about the time period of this image.

Around these central elements are houses, billboards, and trees, none of them new, none of them particularly well kept, and none of them in a clear relationship with each other. It is a scene of an unnamed town in decline, made up of the remains of habitation and commerce that have held on. We also see all the additions and changes that have occurred over time. Three people inhabit this world: a bare-chested man in the parking lot, his back turned to a shopping cart filled with what might be his belongings, and two people by a shipping container parked behind the Redemption Center. They look as lost and used up as everything else in the image.

Here, Crewdson has fabricated a whole world. In that fake scene, a span of time—evidenced by buildings and props from various periods, the wear and tear and additions on those objects, and the people standing around in them—is laid before us. It is a kind of imaginative documentation.

Another tactic to create such a continuous recreation is to mix the real and the projected, altering and extending buildings through the imagery we let appear on them. The work of the American artist Doug Aitken transforms buildings, however fleetingly, into other places and times. In one of his art pieces, called *Song 1* (2018), he projected scenes he had shot at night onto the exterior of the circular Hirschhorn Museum in Washington, DC. The images, all from the Los Angeles area, included a coffee shop, an apartment, an office, a store, the city's port, the inside of factories, parking garages, and the streets near Los Angeles International Airport. Some of these scenes appeared to be set in the 1950s or '60s, while others were in hypermodern environments. All were as dark and artificially lit as the setting in which visitors saw them—the work was only presented at night.

The scenes were also mostly empty, populated only by a few solitary people marooned over a cup of coffee, walking the street, or working away in the vast halls of industry. Now and then, a close-up would narrow the scale, and you would see an old-fashioned reel-to-reel tape deck spinning away, drawing you back to *Song 1*'s soundtrack: "I Only Have Eyes for You," recorded in 1959 by the Flamingos, a trio that now croons the refrain along with actors both famous and unknown in a new version of the old recording. *Song 1* made Los Angeles into a sprawling yet intimate presence. The iconic song, together with the mixture of periods depicted in the scenes, collapsed time as well as space. You were confronted with the essence of LA, but you were still in Washington, DC—or were you?

INSTALLATIONS

All these art projects are restricted by the fact that they can evoke spaces, but they cannot bring us into those spaces. Installation artists, however, can surround us completely with their uncertain scenes. The master of this kind of work is the British artist Mike Nelson. During the first decade of this century, he made labyrinths inside buildings that transported you to other worlds yet were distinctly recognizable. One of the most complete of these was *A Psychic Labyrinth*, constructed inside Manhattan's Chelsea Market while it was under reconstruction in 2007. You entered the building, expecting large, empty spaces and instead found yourself wandering through a warren of shops. There were tattoo parlors and Chinese food stores, as well as offices of uncertain intent, all made by Nelson.

To find your way out, you quickly discovered you had to open doors and enter into some of these shops, encountering environments both banal and unsettling. (Why was there a straitjacket hanging there, and why was an American flag crumpled in one corner, splattered in blood?) You felt your way forward with an increasing sense of dread, especially when you encountered doors that looked exactly like the ones you had just walked through. In the end, you escaped into a large room where, under the still unrenovated ceiling of the market, you encountered a sand dune. Looking back, you realized the whole neighborhood you had struggled your way through was under that artificial beach.

"Beneath the paving stones, the beach," was the slogan the anarchists used when they spearheaded the Paris uprising of 1968. With Nelson, the situation was reversed: below the sand was a world that was rather scarier but also strangely familiar to those who might have experienced the undeveloped parts of New York a few decades earlier. What made that sense of dislocation work so well was Nelson's ability to painstakingly recreate banal spaces—complete with architecture and furniture but also with stains, layers of coverings, cigarette butts, and everything else that made them real—in a way that was just slightly off. You began to question everything about your memory as well as the spaces you encountered.

The Louisville-based duo Jonah Freeman and Justin Lowe created a similarly off-kilter labyrinth during the Art Basel fair in Miami in 2008. Called *Hello Meth Lab with a View*, it occupied the two floors of a duplex unit in a condominium building under construction. The artists brought you into what appeared to be an apartment that had been taken over by a band of illegal drug makers whose lives were as scuzzy as you might imagine from media reports on the trade (the series *Breaking Bad* had not yet made its debut). There was not only garbage and clothes strewn around the installation but also broken toilets and holes in the walls. If you turned your eye from that mess, you could see the water of the bay glistening under the southern Florida sun before you turned back to the actual lab equipment. Then you exited again through a different unfinished corridor.

What made Nelson's and Freeman & Lowe's work so affecting was the combination of the careful craft of making the artists applied to it and a sensibility that found beauty in leftovers, damaged spaces, and what we would usually think of as ugly textures and colors. Inserted moments within that jumble let you escape from the claustrophobia their labyrinths induced. The effect was to combine the opposite of conventional beauty with bits of sublimity that were both terrifying and liberating. Coming out of these installations, you felt as if you had been transported to a completely other, convincing, and yet undefinable space, while never leaving an actual place. The unreal reality was complete.

Other artists have created environments that also took you to strange places, but with a clearer agenda. One of the most ambitious of these was *Tomorrow*, an installation with which the Berlin-based duo Elmgreen & Dragset filled several galleries of London's Victoria and Albert Museum (V&A) in 2014. In this case, the artists presented the installation explicitly as the set for an (as of yet) unproduced play, and even offered visitors a script instead of a gallery guide.

The conceit here was that this was the rather grand, neoclassical apartment of an elderly architect. He had collected a wide variety of artifacts that reflected both his interest in his field and his queer sensibility. The pieces were all part of the V&A's collections. Elmgreen & Dragset used the galleries more or less as they'd found them, just as so many of the architects mentioned above have accepted the sites in which they work, realizing that they were only slightly larger and more formal than the ones you could imagine an old-fashioned successful British architect inhabiting.

Not all the rooms were grand. This architect had an office where files and mementos were piled up on mid-century furniture. It was also clear that the inhabitant had either recently passed away or was relocating: holes in walls indicated that some renovations were taking place, and moving boxes were piled around the rooms. Then, just as you thought the place was uninhabited, you came across a small boy in a school uniform crouched in front of a fireplace, above which hung his portrait. The boy was a sculpture, but a remarkably realistic one, and you were left to wonder whether this was the architect as a young boy or an object of his desire.

Though Elmgreen & Dragset's installations are more organized and less ambivalent in the attitude of making a fictional past, like Nelson and Freeman & Lowe they completely immerse you, suspending the sense of where and when you were before you entered and where you have wound up. They have also focused on queer space as the subject of their work, creating spatial frameworks in which desire remains implicit and suppressed because it must be, seeping out in images on the wall.

VIDEO AND SOUND

There is a minimalism in the best of Elmgreen & Dragset's work that is taken even further by artists who realize that you can transport people with art without building much of anything. One of the best of these is Janet Cardiff, who works mainly with video and sound. One of her early pieces (from 1995) was no more than an audio guide that you picked up on entering a museum. Her recorded voice would guide you toward works of art and start explaining them before interrupting herself with a "Wait, did you see that?" and veering off into a description of someone or something that was not there. Pretty soon you were wandering through a space that didn't correspond to what you were seeing with your own eyes, even as Cardiff would occasionally return to the more standard narrative about the paintings on display. When she did the project for the San Francisco Museum of Modern Art, she extended the technique into video so that the "ghosts" she described would actually appear and you would be led virtually into the remote bowels of the museum where naked people would streak past art handlers moving pieces through the corridors.

Cardiff has also taken her tours into the streets, mixing descriptions of reality with stories about what might have happened in the city or greeting people who are not actually there. She has tested the limit of creating disjunctions with her *Motet* installation (2001), which consists of a circle of audio speakers that play the chanting of Gregorian monks. You stand in the middle of the array or wander in and out of the circle, letting the voices bathe you and hearing overtones that are not actually there. I have experienced the piece in a medieval chapel and also in white-walled museums, and its ability to use sound alone to dissolve whatever surrounds you is remarkable.

Even more minimal, if that is possible, are the installations, or performance pieces, by the artist Tino Sehgal. He creates what he refers to as "constructed situations," a phrase that might stand for much of the work I have been describing.[4] Given the opportunity to fill the main spiraling gallery of New York's Guggenheim Museum in 2010, he added nothing physical. In the middle of the space a man and a woman writhed in an embrace, rising up occasionally to address the audience with the announcement that this was *The Kiss, Tino Sehgal, 2003* (the year the artist first presented the piece). As you then moved up the spiral ramp, you encountered no works of art on the blank walls, which made you look at those surfaces in a new way. Then a small child would take your hand and ask you what you thought of global warming. You would enter into a discussion as you walked one loop of the spiral. Then the child would hand you off to a slightly older child, with the younger one whispering into the older child's ear what you had been talking about so the older child could continue the conversation with you. So it went up the ramp, with you being handed off to a new and older person at each turn, until the last person—in my case a retired filmmaker—told you this was *The Conversation, Tino Seghal, 2010.*

Then there is a Sehgal that consists of a room that is completely dark. After entering, you shuffle around nervously, and then an unseen person begins to sing. Now and then somebody else joins in. After a while, somebody might take your hand and dance with you. As your eyes become accustomed to the dark, shadows of fellow visitors and actors appear, until you are swept up in the movement of people who you can vaguely identify and who, perhaps because of that blur, seem somehow familiar. Sehgal calls the piece *This Variation (2013)*, and it can be created in any space.

All these works have the quality of making different the space that you inhabit, creating narratives out of thin air, and making you disbelieve your own senses, while believing in what you encounter along the way. The works take you into a possible world that is also impossible, a very real environment that is yet completely fake. This is something that good writers, artists, and filmmakers have been doing for a while. The difference here is the uncertainty between what is real and what is not. In a work of written, painted, or filmed art, you know you are experiencing something that somebody made up, no matter how realistic the brushstrokes or how overpowering the imagery. You can look up or away from the book or painting and still be in a real place outside of the world the artist or writer has evoked. Here, there is no escape, at least not immediately. You must accept that you are somewhere else and in a different time period, and either go with that flow or fight your way back to a place you recognize.

Perhaps even more disturbing are those interventions you only see out of the corner of your eye, making you wonder whether you are experiencing "a glitch in the matrix." In 2015 the artist duo Bitnik repainted part of the facade of the House of Electronic Arts, a museum in Basel, Switzerland, to install a piece called *Glitch*. Walk by the building and at first nothing seems to be different—until you notice the bands of gray-and-white paint covering different sections of the walls suddenly shift so that they are out of kilter, then resume their previous course. Look more carefully, and part of the structure also appears to have briefly come out of alignment—even the metal handrail on top of the loading platform has a kink in it. Keep walking and everything will return to normal, but the sense that something is not right remains with you.

It seems to me that our awareness of the play of real and unreal is especially keen when it revolves around the relationship between the found and the fabricated, the made and the remade, and the projected and the real. The fact that it is increasingly difficult to tell any of those categories apart makes it even more important for architects—who manipulate not just ideas or abstractions but our everyday environments—to at least make us aware of the confusion. Some architects do so by arguing for a return to the real: unambiguous materials (as opposed to veneers or stone columns reproduced in plastic), and building methods that eschew machinery of any sort. Others delight fully in the digital, believing that "parametric architecture," as the designer Patrik Schumacher calls it, is our destiny and we should design forms that we can understand as being only possible using the latest technology.

Yet, both positions—between which most of the architecture discipline floats today—leave us with the tremendous waste inherent in new buildings, while also being stuck in the imagery, power structures, and building methods of the past. They also do not allow us to fully use the past—in all its contradictions and continuities—as a living entity that we can inhabit, make our own, and project out into the future.

The kind of fake architecture I am interested in does let us exist in that uncertainty. I frankly do not care too much whether those efforts have the name of art or architecture, whether they are internally consistent or haphazard in their approach, or whether they are wholesale fakes. What matters much more to me is that the architecture lets us experience where and when we are, where we have come from, and where we might be going, while at the same time doubting our own experiences.

CONCLUSION

Like almost all architects, I was trained to look toward monuments of the past as emblems of perfection, ideal and finished in what they represented, even if they were ruins. You couldn't even consider reusing Athens's ancient temple of the Parthenon, the Egyptian or Mesoamerican pyramids, the cathedral of Notre Dame in Paris, or the Lincoln Memorial. You could only admire them, worship them even, and then make new things inspired by or modeled on their forms. Moreover, what you did today was meant to last as long as these monuments have, if not longer, so that someone someday would recognize you as the (ironically unknown) genius designer of a monument.

The reality has long been that the fast pace of change in how we build, use, and interpret our architecture means that structures change almost as soon as they are finished, and sometimes even before. Companies have to redivide their office spaces; people need to add bedrooms to their houses. Olympic stadia, once the games are over, are turned into shopping malls. Although buildings are generally designed for one purpose, if a new owner does not like or need it, the dumpsters quickly fill with the construction material that the original architects had so carefully designed. Then trucks full of new office partitions, marble, wood, or steel show up. Even if a building is considered so beautiful that it is preserved and kept the way it was, the cost of doing so is often exorbitant: a recent and very faithful restoration of a 1960s museum in Berlin cost three to four times as much as constructing the original building.

There is another way to look at the history, and therefore at what is important in architecture. We have long ignored another tradition in the field, one that started not with the making of places of worship or places to bury people or resources but with the weaving together of natural materials into the tents that first housed human beings. We either carried that lightweight structure with us or let its materials go back to nature when we left our campsite. Around the world we can find examples of buildings that are meant to last only as long as we need them and then can either be moved somewhere else or reused. That tradition is kept alive in some religious structures in Japan and West Africa but also in our festivals and fairs. There, the structures can be as grand and impressive as the palaces and cathedrals we usually turn to as models from the past, even if the temporary structures are ephemeral by nature. We should commit to our buildings only as long as they are useful and meaningful, and then do something else with their parts once they no longer serve those purposes.

Architecture can thus be a form of hunting and gathering, in which structures come together for as long as we need them out of what is already on hand. That was the way we lived long ago, but also the way many artists create their work today. In the work of such artists as Theaster Gates and Tyree Guyton, this collage, or assemblage, becomes a very direct model for architecture. We need to learn from this method and adopt it. We need to treat the cities and suburbs around us as resources and mine them. We need to go dumpster diving, finding and revaluing what we have cast off. That also has a social implication: it is not the monuments of rulers we should turn to, but that which was previously not valued. To make this method work, we need to create not only models that can serve as best-practice examples but also catalogs and inventories of existing materials, thereby countering the array of materials available at big-box stores and construction companies' inventories.

Even beyond ephemeral architecture, we are reusing existing structures in beautiful ways, including as houses for single families or small groups of people in Belgium and England, libraries in the Netherlands, as shopping malls in the United States, and as old railroad yards and steel mills in Germany and China. Former industrial landscapes are becoming public places and office buildings are becoming apartments, while lofts, where things were once made, now allow people to make their own lives. Many of these renovations are banal and not as efficient as they should be, but there are enough inspiring examples that can serve as beacons of what is possible: the beautiful reimagination of the boxes in which we live, work, and play.

Our contemporary monuments should be not just temples and tombs but the remains of how we made things, such as factories, and the landscapes that connect us. Our experimental outstations can be houses and small structures, and the most important work we can do in architecture is to create spaces—such as libraries, community centers, and places to shop or just be with other people—that bring us together in a way that is open but informed by history. There are countless examples from around the world—some of which I have pointed to in this volume—that can serve as models for imaginative reuse that is sustainable, open, and beautiful.

For those of you who are architects, I hope you think twice about taking a commission that demands new construction. You should work hard to convince your client that there are plenty of buildings worth renovating. Even if doing so might be more expensive and troublesome in the short-term, our debt to our planet and to our fellow citizens who have been excluded from our buildings and culture would make that extra effort more than worth it. If you do renovate and reuse, do so in a manner that does not tear out or cover over what is there, but that rather makes what remains part of the life of the building today and tomorrow. Do so by revealing and incorporating the fabric of the existing structure as much as possible into the reuse project. Show not just the history of each material or building, but also open them up to interpretation: fragments and collage are better than the appearance of finish. Always use existing materials whenever possible, whether in reuse or—if you must—in new buildings. New buildings can be made by upcycling what exists. Go dumpster diving, or at least shop for existing materials, and think beyond the standard inventories of suppliers of recycled wood, windows, and doors.

For those of you who teach architecture, I hope that the examples in this book have been inspiring and that you will incorporate them in your lessons. Too few of our schools of architecture, landscape architecture, planning, and design incorporate the notion of reuse and recycling into their core curricula. They belong there—not just as adjunct courses for students seeking to specialize.

For those of you who use and enjoy (or are irritated by) architecture, go visit some of the sites I have mentioned. You can also work in your own community to preserve, protect, and promote the reuse of buildings, especially civic ones. Please do not, however, preserve a building, a neighborhood, or a park in amber, but open it up to new uses, users, and interpretations.

It is up to all of us to reuse what we have inherited in an imaginative manner to build a better world.

ACKNOWLEDGMENTS

I would like to thank Virginia Tech for its support of my research on this project, as well as Afrida Afroz Rahman, Frankie Keene, and Tharun Bhalaji for their help as my research assistants, and Professor Bryan Green for his advice. I am also very grateful to all the architects who were so generous with their time in showing me their work, and in particular to Frank Gehry to first awakening me to the idea of imaginative reuse. Thank you also Richard Quittenton for accompanying me on my travels through Europe. I also wish to thank Catherine Tung for her empathetic and precise editing, which has helped shape this volume into a much better narrative.

A NOTE ON SOURCES

This book is based as much as possible on my personal experience with the buildings I discuss. I have traveled extensively through the United States, Central and South America, Europe, and Asia to find what I think are the best examples of imaginative reuse, but of course the result is just a selection of many such projects that I could have included, and I was also not able to see every project I discuss. For basic project information, I relied on the websites of the architects or projects, as well as on the websites that have replaced architecture and design periodicals as primary sources. Among these, the ones that I have found to be the most reliable are:

Archdaily.com
Archello.com
Architectmagazine.com
Architecturalrecord.com
Architizer.com
Archplus.net
Arquine.com
Detail.de
Dezeen.com
Divisare.com
Domusweb.it
Metropolismag.com
Placesjournal.org

In addition, several recent books have collected work such as I discuss and provided theoretical context for them. Among these, I have found the following most pertinent:

Alkemade, Floris, et al. *Rewriting Architecture: 10+1 Actions.* Amsterdam: Valiz, 2020.

Arnaud, Michel. *Cool Is Everywhere: New and Adaptive Design Across America.* New York: Abrams, 2020.

Fannon, David, Michelle Laboy, and Peter Wiederspahn. *The Architecture of Persistence: Designing for Future Use.* London: Taylor and Francis, 2021.

Fernandez Per, Aurora, and Javier Mozas. *Reclaim: Remediate Reuse Recycle.* Vitoria-Gasteiz: A+T Publishers, 2012.

Plevoets, Bie, and Koenraad Van Cleempoel. *Adaptive Reuse of the Built Heritage: Concepts and Cases of an Emerging Discipline.* London: Taylor and Francis, 2019. Kindle.

Robligo, Matteo. *Re-USA: 20 American Stories of Adaptive Reuse.* Berlin: Jovis Verlag, 2017.

Toromanoff, Agata. *Converted: Reinventing Architecture.* Tielt: Lannoo Publishers, 2020.

Wong, Liliane. *Adaptive Reuse: Extending the Lives of Buildings.* Basel: Birkhäuser, 2016.

For theoretical context, the following volumes are useful:

Baker-Brown, Duncan. *Re-Use Atlas: A Designer's Guide to a Circular Economy.* London: RIBA Publishing, 2019.

Baumann, Zygmunt. *Liquid Modernity.* 2nd ed. Cambridge, UK: Polity Press, 2012.

Benjamin, David, ed. *Embodied Design Energy: Making Architecture Between Metrics and Narratives.* Zürich: Lars Müller Publishers, 2017.

Bhabha, Homi K. *The Location of Culture.* London: Taylor and Francis, 2012.

Cairns, Stephen, and Jane M. Jacobs. *Buildings Must Die: A Perverse View of Architecture.* Cambridge, MA: MIT Press, 2014.

Franklin, Kate, and Caroline Till. *Radical Matter: Rethinking Materials for a Sustainable Future.* London: Thames & Hudson, 2018.

Groys, Boris. *On the New.* Translated by G. M. Goshgarian. 2nd ed. London: Verso Books, 2014.

Kaufman, Ned. *Place, Race, and Story: Essays on the Past and Future of Historic Preservation.* London: Routledge, 2009.

Kirkham-Lewitt, Isabelle, and Joanna Joseph, eds. *Sketches on Everlasting Plastics.* New York: Columbia Books on Architecture and the City, 2023.

Koolhaas, Rem, and Jorge Otero-Pailos. *Preservation Is Overtaking Us.* New York: Columbia University Graduate School of Architecture, Planning, and Preservation, 2009.

Lears, T. J. Jackson. *No Place of Grace: Antimodernism and the Transformation of American Culture, 1880–1920.* 2nd ed. Chicago: University of Chicago Press, 1994.

Mayes, Thompson M. *Why Old Places Matter: How Historic Places Affect Our Identity and Well-Being.* Lanham: Rowman & Littlefield, 2018.

Osseo-Asare, D.K., and Yasmine Abbas. "Waste." *AA Files* 76 (2019): 179–83.

Rogers, Louis, ed. *Material Cultures: Material Reform.* London: MACK, 2022.

Stockhammer, Daniel, eds. *Upcycling: Reuse and Repurposing as a Design Principle in Architecture.* Triest: University of Liechtenstein, 2020.

Stone, Sally. *Undoing Buildings: Adaptive Reuse and Cultural Memory.* New York: Routledge, 2020.

Wong, Liliane. *Adaptive Reuse: Extending the Lives of Buildings.* Basel: Birkhäuser, 2016.

For historical context, see:

Nagel, Alexander, and Christopher S. Wood. *Anachronic Renaissance.* New York: Zone Books, 2010. Kindle.

Nietzsche, Friedrich. "On the Utility and Liability of History for Life." In *The Nietzsche Reader*, edited by Keith Ansell Pearson and Duncan Large, 124–41. Oxford: Blackwell, 2008.

Page, Randall Mason. *Giving Preservation a History: Histories of Historic Preservation in the United States.* New York: Routledge, 2004.

Riegl, Alois. "The Modern Cult of Monuments: Its Character and Its Origin." In *Historic Preservation Theory: An Anthology; Readings from the 18th to the 21st Century,* edited by Jorge Otero-Pailos, 118–22. New York: Design Books, 2022.

Roark, Ryan. "The Afterlife of Dying Buildings: Ruskin and Preservation in the Twenty-First Century." The Courtauld, https://courtauld.ac.uk/research/research-resources/publications/courtauld-books-online/ruskins-ecologies-figures-of-relation-from-modern-painters-to-the-storm-cloud/14-the-afterlife-of-dying-buildings-ruskin-and-preservation-in-the-twenty-first-century-ryan-roark, accessed December 13, 2023.

Rossi, Aldo. *A Scientific Autobiography.* Translated by Lawrence Venuti. Cambridge, MA: MIT Press, 1981.

Ruskin, John. "The Lamp of Memory." In *The Seven Lamps of Architecture,* 176–98. New York: Dover Editions, 1989. First published 1849 by Smith, Elder & Co. (London).

Stewart, Susan. *The Ruins Lesson: Meaning and Material in Western Culture.* Chicago: University of Chicago Press, 2020.

Tyler, Norman, Ilee R. Tyler, and Ted J. Ligibel. *Historic Preservation: An Introduction to Its History, Principles, and Practice.* 3rd ed. New York: W. W. Norton, 2018.

Viollet-le-Duc, Eugene-Emmanuel. *The Architectural Theory of Viollet-le-Duc: Readings and Commentary.* Edited by M. F. Hearn. Cambridge, MA: MIT Press, 1990.

NOTES

INTRODUCTION

1. "Circular Economy Action Plan," Environment, European Commission, https://environment.ec.europa.eu/strategy/circular-economy-action-plan_en, accessed December 13, 2023.

2. Liliane Wong, *Adaptive Reuse: Extending the Lives of Buildings* (Basel: Birkhäuser, 2016), 38, Kindle.

3. Wong, *Adaptive Reuse*, 85.

4. Svetlana Boym, *The Future of Nostalgia* (New York: Basic Books, 2008), 49, Kindle.

5. David Fannon, Michelle Laboy, and Peter Wiederspahn, *The Architecture of Persistence: Designing for Future Use* (London: Taylor and Francis, 2021), 18, Kindle.

6. Fannon, Laboy, and Wiederspahn, *The Architecture of Persistence*, 238.

7. Rem Koolhaas, "Preservation Is Overtaking Us," in Jorge Otero-Pailos, *Preservation Is Overtaking Us* (New York: Columbia University Graduate School of Architecture, Planning, and Preservation, 2009), n.p.

8. Koolhaas, "Preservation Is Overtaking Us."

9. Robert Venturi, *Complexity and Contradiction in Architecture* (New York: Museum of Modern Art, 1966), 16.

10. Daniel Stockhammer, ed., introduction to *Upcycling: Reuse and Repurposing as a Design Principle in Architecture* (Trieste: University of Liechtenstein, 2020), 14–33, 19.

11. Andreas Hild, preface to Stockhammer, *Upcycling*, 9–13, 9.

CHAPTER I: THE ARCHITECTURE OF REMAINS

1. The statistics on this vary widely, but the most reliable come from the World Bank, https://www.worldbank.org/en/country/mic/overview, accessed February 25, 2024.

2. Eugene-Emmanuel Viollet-le-Duc, "Restoration," from *Dictionnaire Raisonné*, vol. 8, in *The Architectural Theory of Viollet-le-Duc: Readings and Commentary*, ed. M. F. Hearn (Cambridge: MIT Press, 1990), 269–79, 270.

3. Viollet-le-Duc, "Restoration," 275.

4. Eugene-Emmanuel Viollet-le-Duc, "Notre Dame: 'Report Addressed to the Minister of Justice and Religious Rites,'" in Hearn, *The Architectural Theory of Viollet-le-Duc*, 279–88, 280.

5. John Ruskin, "The Lamp of Memory," in *The Seven Lamps of Architecture* (1849); (New York: Dover Editions, 1989), 176–98, 178.

6. Ruskin, "The Lamp of Memory," 194.

7. William Morris, Philip Webb et al., *The Manifesto of the Society for the Preservation of Ancient Buildings*, 1877, https://www.spab.org.uk/about-us/spab-manifesto, accessed January 19, 2024 (printed source not available; this is the statement verified by the organization that published it originally).

8. Alois Riegl, "The Modern Cult of Monuments: Its Character and Its Origin," in *Historic Preservation Theory: An Anthology; Readings from the 18th to the 21st Century*, ed. Jorge Otero-Pailos (New York: Design Books, 2022), 118–22.

9. Riegl, "The Modern Cult of Monuments," 118.

10. *International Charter for the Conversation and Restoration of Monuments and Sites*, passed at the Second International Congress of Architects and Technicians of Historic Monuments, Venice, 1964. It is maintained and published by the International Council on Monuments and sites (ICOMOS) at https://www.icomos.org.

11. Riegl, "The Modern Cult of Monuments," 118.

12. T. J. Jackson Lears, *No Place of Grace: Antimodernism and the Transformation of American Culture, 1880–1920* (Chicago: University of Chicago Press, 1983).

CHAPTER 2: DUMPSTER DIVING AND URBAN MINING

1. "Circular Economy Action Plan," European Commission, https://environment.ec.europa.eu/strategy/circular-economy-action-plan_en, accessed March 2, 2024.

2. "About Us," RotorDC, https://rotordc.com/aboutus-1, accessed January 23, 2024.

3. BlueCity, https://www.bluecity.nl/en, accessed March 13, 2024.

4. Paula Antonelli, "Nothing Cooler Than Dry," in *Droog Design: Spirit of the Nineties*, ed. Renny Ramakers, Gijs Bakker (Rotterdam: 010 Publishers, 1998), 12–15, 14.

5. "Before Yesterday We Could Fly: An Afrofuturist Period Room," https://www.metmuseum.org/exhibitions/afrofuturist-period-room, accessed December 13, 2023.

CHAPTER 3: EPHEMERAL ARCHITECTURE

1. Rahul Mehrotra and Felipe Vera, "Learning from the Pop-Up Megacity: Reflections on Reversibility and Openness," in *Kumbh Mela: Mapping the Ephemeral Megacity*, ed. Mehrotra and Vera (New Delhi: Niyogi Books, 2015), 392–406, 399.

2. Mehrotra and Vera, "Learning from the Pop-Up Megacity," 397.

3. Mehrotra and Vera, "Learning from the Pop-Up Megacity," 403.

4. Robert Reich, *The Work of Nations: Preparing Ourselves for 21st-Century Capitalism* (New York: Alfred A. Knopf, 1991).

CHAPTER 4: GHOST ARCHITECTURE

1. Constance M. Lewallen, *500 Capp Street: David Ireland's House* (Berkeley: University of California Press, 2015), 14, 41.

2. Lewallen, *500 Capp Street.*

CHAPTER 5: SQUATTING, INSTALLING, AND ACTIVATING

1. Theaster Gates, "Accumulations," in *12 Ballads for Huguenot House* (Chicago: Museum of Contemporary Art, 2012), 70.

2. Amy Starecheski, *Ours to Lose: When Squatters Became Homeowners in New York City* (University of Chicago Press, 2016), p. 56.

3. Starecheski, *Ours to Lose*, 241.

4. Starecheski, *Ours to Lose*, 246.

5. Marion E. Jackson, "Trickster in the City," in Tyree Guyton, *Connecting the Dots: Tyree Guyton's Heidelberg Project* (Detroit: Wayne State University Press, 2007), 23–38, 24.

6. Tyree Guyton, "From the Artist," in Guyton, *Connecting the Dots*, v–vii, vii.

CHAPTER 6: HOUSING

1. "Carles Oliver Refurbishes Rustic Home in Mallorca to Preserve Its Material Heritage," DesignBoom, https://www.designboom.com/architecture/carles-oliver-san-miquel-19-palma-mallorca-10-25-2017, accessed March 2, 2024.

2. Jan Jongert in conversation with the author, May 25, 2023.

3. Jongert in conversation with the author, May 23, 2023.

4. Jongert in conversation with the author, May 23, 2023.

CHAPTER 8: REUSING THE LANDSCAPE

1. The libretto, written by Anne Carson and Claudia Rankine, was published by the Mile Long Opera at http://milelongopera.com/MLO_program_libretto.pdf.

2. The quotes in this paragraph are lines written by Carson.

3. Reich discusses this notion most fully in his *The Work of Nations: Preparing Ourselves for 21st-Century Capitalism* (New York: Alfred A. Knopf, 1991).

4. Yu Kongjian in conversation with the author, Beijing, November 14, 2019.

5. Oliver Wainwright, "China's Rural Revolution: The Architects Rescuing Its Villages from Oblivion," *The Guardian*, March 24, 2021, https://www.theguardian.com/artanddesign/2021/mar/24/chinas-rural-revolution-architects-rescuing-villages-oblivion-tofu-rice-wine-lotus-tea.

CHAPTER 9: THE WORLD OF IMAGINATIVE REUSE

1. Pelle Rademakers in conversation with the author, May 28, 2023.

2. Quoted in Natasha Levy, "The Standard Completes New Hotel in London," *Dezeen*, May 10, 2019, https://www.dezeen.com/2019/05/10/the-standard-london-uk-shawn-hausman-design/.

CHAPTER 10: THE MONUMENTS OF IMAGINATIVE REUSE

1. Statistics vary a great deal on this number, depending on what is counted as office space and what is seen to be full utilization. For a conservative evaluation, see "The Impact on Real Estate," McKinsey & Company, https://www.mckinsey.com/mgi/our-research/empty-spaces-and-hybrid-places-chapter-2, accessed March 2, 2024.

CHAPTER 11: FAKING IT

1. Rem Koolhaas, “Prada Foundation,” https://www.oma.com/projects/fondazione-prada, accessed January 19, 2024.

2. Tom Ravenscroft, “Amin Taha Creates Distorted Replica of 19th-Century London Terrace Block,” *Dezeen*, April 1, 2019, https://www.dezeen.com/2019/04/01/amin-taha-groupwork-168-upper-street-london/.

3. Ravenscroft, “Amin Taha Creates Distorted Replica of 19th-Century London Terrace Block.

4. Lauren Collins, “The Question Artist,” *New Yorker*, August 6, 2012, https://www.newyorker.com/magazine/2012/08/06/the-question-artist.

INDEX